Lifespan Extension and longevity secrets through Alchemy, SIRT1 & SOX9 genes and the Maillard reaction

A researcher's notebook on the latest anti-aging substances and Alchemical Longevity Formulas

SCOTT RAUVERS

Read the first 3 chapters of this book FREE by visiting www.scott-rauvers.com/sox9.html

ISBN 978-0-646-87302-2

9 780646 873022 >

ISBN: 978-0-646-88359-5

DEDICATION

This book is dedicated to Colleen Lashway on the beautiful Hawaiian island of Oahu, for her years of service and dedication to the Hawaii Kai Public Library, and for her loyal commitment to the community of Hawaii Kai.

Also by Scott Rauvers................

- Secret Strategies and Techniques the Pros use for Reversing Aging.
- Reverse Aging Naturally. Alchemy and Ayurveda Longevity Anti-aging Secrets.
- Medicinal and Edible Healing Plants of Australia's Sunshine Coast and the Noosa Heads Wilderness
- Reversing Osteoporosis/ Insomnia using newly uncovered Secrets of Bone Breathing.
- Eternal Youth via Tao Te Ching. Longevity Secrets via Universal Energy.
- The Vegetarian's Guide to Longevity via Gene Therapy and Raw Foods.
- The Complete Guide to Natural Toothache Remedies and Re-mineralization.
- Foods, Herbs and Pharmaceuticals that Extend Lifespan
- Anti Aging Nutrition Secrets. The Fountain of Youth Seeker's Guide to Longevity-
- A Centurion's Toolbox for Longevity Living Beyond 120 - 2nd Edition
- Living Healthy Beyond 120, A Centurion's Plan for Longevity
- The Official Guidebook of How to Make Tinctures and Alchemy Spagyric Formulas

All correspondence welcomed

Scott Rauvers
P.O. Box 44
Fortitude Valley
Brisbane, QLD 4006
Australia

Over the past decade I have saved all the very best energy and health formulas from my suite of anti-aging books and put them on my website www.scott-rauvers.com. These formulas can be downloaded and printed out at your convenience for free and at no charge. Please remember that some herbs may interact with prescription medication and always trust your intuition when it comes to dosage.

CONTENTS

INTRODUCTION

During my freshman year in College I was first introduced to the science of aging when I took Gerontology as my major. During this time, I always pondered that there must already exist an antidote for people to live past 100 on a regular basis, and in good health. For example in the book of Genesis in the bible Methuselah was reported to have lived 969 years, Mahalalel 895 years, Lamech 777 years, Enosh 905 years, Kenan 910 years and Jared 962 years. Today more people are living longer than at any other time in history (with the exception of the biblical era). For example 92 year old William Shatner is still doing interviews and writing books and Carmen Dell'Orefice who is 91, is the world's oldest working supermodel. It is hoped the reader of this book joins the growing number of centenarians and supercentenarians. So why are we humans as a species living shorter lives, when at an advanced age of industrialization we should be living longer?

I have also been very pleased to see that many of the theories that I put forth in my 2012 book titled: Solar Flares and Their Effects Upon Human Behavior and Health, which stated that above average solar activity negatively impacts health and well-being, especially in the most vulnerable populations, that since then a number of published studies have come forward confirming my hypothesis. I am also so happy to see that much of the anti-aging discoveries I published years ago are now starting to surface in the published anti-aging literature, giving confirmation to many of my early theories on anti-aging. Especially my early research into probiotics, of whose benefits are now starting to gain mainstream recognition.

It has only been during the last 3 years that technology has made it possible to explore the anti-aging substances in thousands, if not soon to be millions of plant species and their abilty to reduce free radicals and exhibit anti-aging properties. This book contains many of those studies with almost 900 cited references. This has resulted in this book having many major discoveries and breakthroughs. This book is meant for people who have a general understanding of the red powder of the Philosopher's Stone as well as the general terms regarding anti-aging. Endurance athletes can benefit from the information contained in this book for rapid recovery from heavy exercise or endurance sports.

As Lockheed Senior Engineer Boyd Bushman once said, "Follow the data to find the discoveries". The groundbreaking discoveries shown throughout this book are just that, from following the data.

C-reactive protein levels (CRP)

When your body is under prolonged stress and is low in specific nutrients, your liver makes up for a lack of nutrients by trying to fight the stress. Your liver than begins to experience chronic inflammation. However just before this inflammation occurs, there is an increase in C-reactive protein levels in your blood.

For example, a recent July 2022 research study (1) examined the effects of above average solar activity on the elderly in Boston, Massachusetts. The study involved 742 people and took place between May 2000 and December 2017. The average age of the participants was 73 years. The researchers found that changes in sunspot and geomagnetic activity caused their T-cells, which regulate the immune system, to become over-active. (T-cells help fight off disease and viruses). The study went on to state that above average solar weather conditions affected their body's C-reactive protein levels (CRP). The researchers discovered a significant increase in CRP levels depending upon sunspot activity. However CRP levels were not affected by increased geomagnetic activity on the earth, only the sunspot activity. The study also found that these effects were similar to that of air pollution. What earth's geomagnetic activity did affect regarding the participant's health was their fibrinogen levels, which was not affected by sunspot activity. Fibrinogen helps regulate blood clots (helps stop bleeding) and healthy levels show a person is less at risk for cardiovascular disease.

Another study (2) found that for each increase in CRP levels, there is a 1.4 year reduction in lifespan in people aged 57 years or more. This study shows that CRP levels can act as a way to predict longer-term survival in middle aged men. Other published studies regarding solar activity and health include, Geomagnetic disturbances driven by solar activity enhance total and cardiovascular mortality risk in 263 U.S. cities (2019. Carolina Leticia Zilli Vieira et al) and Revealing the relationship between solar activity and COVID-19 (Mohammad Hossein Nasirpour et al. 2021).

Being one of the world's most respected experts on longevity nutrition for over a decade, I have discovered from my writing of 5 books on anti-aging which span thousands of pages, that the secret to longevity consists of a 2-fold process. 1 - Identifying longevity substances that exist naturally in your environment (or the right supplements). 2 - Having a calm mind and heart.

The environment you live in also relates to your longevity diet. If you lived in an area where there is lots of rain, you would eat more dry foods such as adzuki beans. If you lived in a dry region, you would eat more damp foods such as yogurt. Also if you were a person that ate a lot of raw foods, you would not live in an area with cold winters, but would be more suited to warmer climates. You can also do an Internet search using the term - **TCM dry foods** - to find numerous charts and articles regarding Traditional Chinese Medicine and damp and dry foods.

When author / scientist Greg Bradden travelled to some of the most remote places on earth and interviewed the longest living people (3) he discovered that they all stated that one of the reasons they lived so long was because they had calm hearts and minds. In other words, they did not allow their emotions to get the better of them or see small minor events in their lives as major obstacles or allow them to cause them to get frustrated. They kept things that happened in their lives in healthy perspective. One way to have a calm heart to relax. For example, studies show (4) that practicing meditation short term creates considerable effects on brain energy, which helps reduce anxiety and depression.

Any experienced self-help coaching professional or a person who has been a motivational speaker for decades knows that between 1% and 3% of people aged 60 or older are very independently wealthy. They also know that independent self-wealth is a state of mind which stems comes from having the right mindset. I also came across this information when I wrote 3 of my self-improvement books (Learn to Create Wealth and Manifest Infinite Financial Abundance, Modern and Secret Teachings of Eternal Wisdom, Peace, Abundance and Prosperity and Secret Teachings for Manifesting Prosperity using Infinite Spirit (which you can read for free at scott-rauvers.com). This trend also appears to take place in centenarians (a person who lives past 100 years of age). According to research by Boston University (5), out of the entire population of the United States, centenarians existed at the rate of 0.27% (2021), with the highest numbers of centenarians found to be living in Hawaii and Connecticut. So in order to live past 100, it requires the right mindset and a calm heart is very likely the key.

A recent 2022 research study published by Columbia University (6) stated that getting good sleep is vital for your heart. They state that sleep impacts the heart and that not enough sleep affects the heart by influencing our decisions about food. Their research showed that lack of sleep, or not sleeping well, leads to food cravings which cause one to be more likely to eat foods high in saturated fat and sugar. This could be why research studies have proven that people who have good sleeping habits have up to a 35% lower chance of dying from heart syndrome (7). Indeed studies show (8) that a positive relationship exists between sleep quality and lifespan in centenarians, including people aged 80 or more, with the optimal range being between 7 and 9 hours sleep each night.

How Plants can Extend Human Lifespan

The purpose of this bfook is to give you, the reader, the best and latest research regarding natural plants and their associated extracts scientifically proven to extend lifespan. The use of plants for healing and longevity is a respected ancient tradition that is now much older than today's modern medical science of today [9]. It is estimated that today there are approximately 80,000 plants used worldwide for healing and longevity medicine [10]. It is a fact in the anti-aging literature that the most successful anti-aging herbs and substances also treat, eliminate or prevent cataracts. For example, the substances baicalein, forskolin, hesperidin, resveratrol and ginsenoside all lower intraocular pressure [11]. Resveratrol is a substance well known for its lifespan extention effects. A research study published in April 2016 titled: Effect of the Resveratrol Rice DJ526 on Longevity, published by Saidul Islam and colleagues found that the resveratrol rice DJ526 significantly extended the lifespan of fruit flies up to 41.4%. This has since been confirmed in follow-up studies (The Resveratrol Rice DJ526 Callus Significantly Increases Lifespan. Mousumee Khan et al. Apr 2019).

Mystery Solved. Why the Bristlecone Pine tree lives for thousands of years

The Bristlecone pine has been rigorously documented to live up to 5,000 years [12]. Many researchers and scientists, including myself for many years, were perplexed as to what actual substance was making these trees live thousands of years. Just as the famed Chinese immortals were reported to live high up in the mountains, the Bristlecone pine tree lives high up in the mountains in an environment that has scathing winds, acidic soil and long cold winters [13]. This cool, dry environment with low rainfall means it exists in an environment that has high hydrophobicity, where dust and dirt cannot cling and adhere to surfaces for long periods of time. What is interesting is the Bristlecone pine tree grows in dry regions that have a soil content high in dolomite. Dolomite is abundant in calcium containing 40% magnesium carbonate and 60% calcium carbonate (14).

A clinical trial (14) looked administering the mineral supplement (Aquamin) which is composed of the red algae Lithothamnion corallioides, which contains 85% calcium and 2.4% magnesium to volunteers diagnosed with severe osteoarthritis of the knees. The clinical trail discovered that the participants who took Aquamin exhibited significant improvements in their 6 minute walking sessions. There were no significant differences observed in the walking distances in the placebo group and as will be explained later in this book an over-expression of the SOX9 gene is used to heal many arthritis disorders. It is interesting to note that the gene level of SOX9 is significantly down-regulated in people diagnosed with osteoarthritis compared to people who have not been diagnosed with osteoarthritis (15).

I hypothesize that the reason this pine tree lives so long is because the harsh environment forces it to produce substances to protect itself, which in turn creates substances that reduce the aging process. Evidence to support this hypothesis is the fact that one of the most powerful anti-aging substances, Resveratrol, is made in over 70 fplant species as a response to stress in its environment. (16). This defense behavior is similar to the process where a sheep is injected with just enough snake poison to not kill the sheep, but to allow the sheep to produce antibodies which are used to treat snake bite.

Phytoalexins

Resveratrol just happens to be a phytoalexin (*stilbene*), which possesses very promising antioxidant properties (17). Antioxidant effects take place when a substance behaves as a hydrogen donor and attaches itself to a free radical, sacrificing itself in the process, so that it may neutralize the free radical.

Phytoalexins slowly accumulate in plants as a byproduct of

their resistance to adverse weather conditions, parasites, fungal attacks, UV radiation, chemicals and general stress in the plant. As just stated, Resveratrol is made by over 70 plant species in response to stressful situations (18), the Bristlecone pine tree must be making a type of Phytoalexin. Research shows (19) that the most abundant substances in the Bristlecone pine are Pinene, Limonene, Phellandrene, Terpinolene, Camphene and Myrcene as well some unidentified substances. Research also states (20) that the Bristlecone pine also contains an abundance of Carene and that Carene has a sleep enhancing effect, improving sleep duration.

The plant Bupleurum gibraltaricum contains 33% Carene (783). When researchers induced swollen inflamed feet in mice, they found that after applying the essential oil of Bupleurum gibraltaricum, that it exhibited considerable anti-inflammatory activity. The researchers concluded this reduction in swelling was due to the carene (Pharmacological activity of the essential oil of Bupleurum gibraltaricum: Anti-inflammatory activity. M.A. Ocete et al. May 1989). This is one of the few emerging studies that show Carene has been used to resolve a health issue.

In the book titled: **Xylem Monoterpenes of Pines: Distribution, Variation, Genetics, Function**, written by Richard H. Smith, he states that the Bristlecone pine trees in eastern Colorado and the Northern regions of New Mexico and Arizona contained between 1% and 5% a-pinene and 75% to 85% carene. Another region that contained 46% a-pinene and 39% carene was located at Spring Mountain Nevada; Tamarack Canyon, California and White Mountains California, with another region at Humphrey's Peak in Arizona. Also Bristlecone pine trees near Mt. Shasta California were found to contain up to 49% carene. His book goes into much more detail, including maps and locations of the regions of the trees. Highly recommended reading. Carene is also emitted by plants as a natural defense against insects (21). Extremely high levels of Carene are emitted by the Sitka spruce tree, whose lifespan is between 700 and 800 years, in order to ward off weevil attacks (22).

A fungus that grows on the bottom of the ocean called

Eutypella scoparia also happens to contain carene [(23)] and this fungus has been found to exhibit powerful natural anti-inflammatory activity. The immortal jellyfish, Turritopsis medusa, sinks to the bottom of the ocean floor where it becomes a gelatinous blob, than over the next few days it becomes a polyp, which allows it to regenerate itself [(24)]. Further research may find that this jellyfish specifically picks locations where the fungus Eutypella scoparia is present when rejuvenating itself.

Carene Levels in Centenarians

Interestingly carene is found in significant amounts in centenarians [(25)] and is associated with healthy aging and longevity. Other substances found in centenarians include 1H-indole, 5 methyl, dimethyl trisulfide and formic acid butyl ester, all of which are anti-aging substances, with carene and 1H-indole, 5 methyl being the most abundant. Lower levels of these were found in younger people, most likely because bodies that are older use these substances to fight age related inflammation. 1H-indole is a substance that is produced by the good bacteria in your stomach (a probiotic) [(26)].

Plants that contain Carene as a major constituent -

Carene levels vary in plants and trees due to geographic region which can be affected by sun and winds, the soil type, elevation and growing conditions. Many of the plants and trees shown below are also abundant in pinene and limonene.

Black Pepper (Piper nigrum) [(27)] - - Leaf oil of Cedrus deodara [(28)] (Roxb. ex D.Don) G. Don (Pinaceae). The Deodar Cedar tree can live as long as 1,000 years and it is used to fight leukemia. - - Korean Pine trees [(28)]. - - Mediterranean cypress [(29)] (Cupressus sempervirens L.) (Cupressaceae), which is also used to fight leukemia. Some species of the Mediterranean cypress can live to over 1000 years - - Greek Juniper [(30)] (Juniperus excela M.Bieb) (Cupressaceae). Greek Juniper showed that in low doses, (10 ppm), that it enhanced the lifespan of C. elegans by 18.54% [(31)]. What was also interesting in the study was that the worms exhibited a 30.40% longer survival rate when placed under thermal stress, compared to the control group. Also they expressed elevated levels of SOD-3 (39.49%) and GST-4 (glutathione) (25.13%). The glutathione gene is a gene that helps the body remove environmental pollutants [(31)] - - Pinus roxburghii Sarg. (Pinaceae) [(32)] which is also used to fight colon cancer. P. roxburghii trees can live as long as 123 years or more.

Alpha Brainwaves and Pinene

When looking at brainwave frequencies over the course of a lifetime between the ages of 3 and 7, theta brainwaves are the most dominant, with alpha starting to increase after age 7 [(33)]. Between 11 and 12 years of age alpha brainwaves peak, than begin to decrease until age 40 [(34)]. Between the ages 60 and 90, there is a major decrease in fast alpha brainwave activity [(35)] and in older people with dementia and people diagnosed with psychiatric diseases, alpha brainwaves will turn into theta waves [(36) (37)].

In a clinical trial involving 10 men and 10 women that inhaled α-pinene [(38)], women experienced significant increases in their alpha waves. When they inhaled β-pinene, they also experienced increases in alpha brainwaves, however the men experienced a decrease of alpha waves. However alpha waves were increased when they inhaled the a β-pinene scent. Another study [(39)] that tested participants inhaling Peppermint essential oil, which contains 15.99% Limonene and 7.71% Pinene, found increased alpha wave activity in the participants' prefrontal cortex region of the brain (up to 31%) when they looked at white pictures. Also when worms were exposed to the scent of Pinene, they showed better resistance to heat stress [(40)].

Interestingly when the Great Basin Bristlecone pine is exposed to higher temperatures, there is a significant decline in α-pinene/limonene levels ($P < 0.001$, $R2 = 0.414$), as well as α-pinene/3-carene levels [(41)]. I hypothesize that the pinene and carene are acting as heat stress reduction mechanisms, which is why they decline in the presence of heat stress. This would also mean that methods the reduce heat stress may be one of the most powerful anti-aging mechanisms.

A little later on we shall explore in great detail why coffee is one of the most powerful anti-aging foods. As an interesting side note, both the words coffee and carene both start and end with the letters C & E. Also the words carene and coffee both contain two E's and each name consits of six letters.

The key is the type of coffee and how long the beans are roasted. When Luwak Green Arabica Coffee is fermented, the most significant compounds produced are α-pinene, furfural and 3-carene [42], and carene is found in Schinus terebinthifolia Raddi [43], which is used to treat respiratory diseases as well significantly reduce blood pressure due to its ability to lower oxidative stress [44]. Carene is also found in lime peel [45] and cinnamon [46].

Chapter 1
Exploring the anti-aging substances in roasted coffee

So now let's explore the reactions that take place in foods when they are heated starting with coffee. Roasting coffee beans will enhance salicylic and caffeic acid levels [50]. Interestingly, dark-roasted coffee contains significantly ($p < 0.0001$) higher salicylic acid levels, compared to medium and lightly roasted coffee beans. As roasting time increases, salicylic acid levels will rise. This has been confirmed by Pelvan et al. (2018). Higher temperatures also boost salicylic acid levels and coffee that experiences short brewing times shows higher levels of salicylic acid [50].

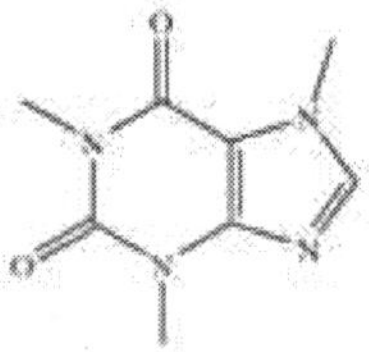

Caffeine

Chlorogenic acid is found in coffee, with concentrations as high as 70 to 350 mg per cup [51], with a negative correlation existing between chlorogenic acid levels and the degree of roasting, with light roast and green coffee having the highest amounts. Regarding country, coffee from Ethiopia has the highest levels of chlorogenic acid followed by Columbia. Columbian coffee also has the best roasting ability [51]. I hypothesize that the salicylic acid in coffee is enhancing the bioavailability of the chlorogenic acid. This could be why numerous studies show coffee extends lifespan. In some studies showing increases in lifespan up to 52% [52] [53] [54]. This means other bio-enhancers may enhance the absorption of chlorogenic acid. Further studies are needed to confirm this hypothesis.

Further Reading

The effect of time, roasting temperature, and grind size on caffeine and chlorogenic acid concentrations in cold brew coffee. Fuller M., Rao N.Z. Sci. Rep. 2017;7:17979. doi: 10.1038/s41598-017-18247-4.

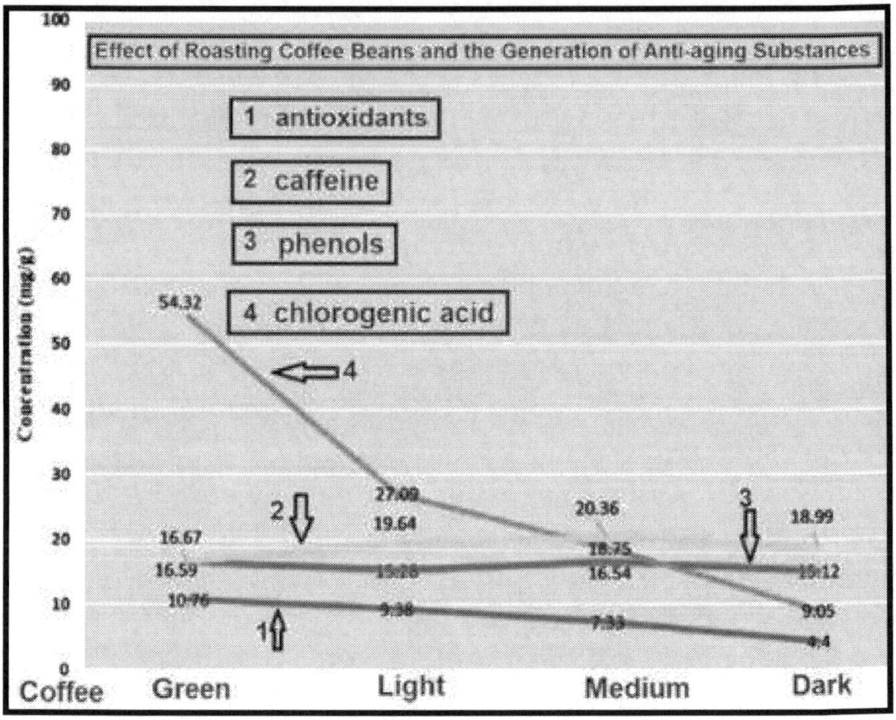

Above Graph Courtesy of: Quantification of Caffeine and Chlorogenic Acid in Green and Roasted Coffee Samples Using HPLC-DAD and Evaluation of the Effect of Degree of Roasting on Their Levels. Shady Awwad, et al. Dec 2021.

What does Bioavailability Mean?

In order to carry out their maximum effectiveness, substances in foods must be fully released from the foods during the digestion phases in the gastrointestinal tract or colon to make them made fully absorbable (Heaney, 2001). If this properly takes place, they are then transported into the body's bloodstream and into systemic circulation throughout the body where they are used for numerous metabolic functions (55) (Wood, 2005). In many cases, the more hydrophilic the substance, the better its bioavailability. A patent titled: Hydrophilic matrix beadlet compositions with enhanced bioavailability (#WO2015173603A1), uses such a composition to enhance the bioavailability of fats and a Sept 2020 research study titled: Direct Conjugation of Resveratrol on Hydrophilic Gold Nanoparticles, published by Iole Venditti and colleagues, used gold nanoparticles created from cysteine and citrate to enhance the bioavailability of resveratrol due to the strongly hydrophilic composition of the nanoparticles.

Regarding plants, the plant species Lonicerae Japonicae Flos, contains considerable levels of chlorogenic acid (56) and an extract of Lonicerae Japonicae Flos has been shown to extend lifespan in worms up to 21.8% (57). It also significantly enhanced the gene bec-1, the gene responsible for normal aging and neurodevelopment and it also reduced toxins. The authors of the study concluded that Lonicera japonica exhibits major implications for promoting healthy aging and to treat age-related diseases.

What do Amino acids and Meteorites have in common?

You are reading one of the few books that explores modern day science to connect the dots of alchemy. So let's start connecting some of the dots. Let's first examine what substances are produced by meteorites as they enter earth's atmosphere.

When a meteorite is travelling through space, its temperature ranges from 100°C to 400°C (58). However as soon as that same meteorite enters earth's atmosphere for its brief 10 to 15 second journey, this mass hurtling through earth's atmosphere will heat up to over 1,800 degrees Celsius (a super Maillard reaction, which we shall explore in more detail later on). If you heat glycine with alumina to 240 degrees Celsius, it will generate the amino acids alanine, alpha-aminobutyric acid, norvaline, norleucine, sarcosine, ethylglycine, methylalanine and ethylalanine (59). These just happen to be the same amino acids found in meteorites (60). These can also be synthesized in the lab through the application of an electric current through water, methane, nitrogen and ammonia (61). Therefore, we have a solid basis to support the hypothesis that meteorites bring life to planets throughout our universe via a cosmic Maillard reaction.

This has been speculated as far back as the 5th century BCE, when Greek philosopher Anaxagoras coined the term Panspermia, which means life on earth came from chemicals in outer space, which initiate life as soon as they encounter the right environment.

During orbital experiments by ERA, BIOPAN, EXOSTACK and EXPOSE (62) the data showed that spores of B. subtilis were killed if they became exposed to the full environment of outer space; being killed in a matter of seconds. However if they were shielded against solar radiation, they were able to survive embedded in clay or meteorite powder. Researchers from both the University of Hawaii and the University of California, Berkeley, (63) (64) concluded that under the right conditions, outer space allows for the creation of amino acids.

Could the Philosopher's Stone be a matter of separating amino acids from a substance than using the right processing methods to create a compound that becomes not only super bioavailable, but that has transmutation properties?

Meteorites Create Gold

A research study published by the University of Bristol stated that gold and **platinum** came to earth from meteorites. The lead researchers in the study, Willbold and Elliott, stated that Earth's gold was the result of years of meteorite bombardment. As the gold-laden meteorites slowly became stirred into Earth's mantle via convection, geological formation created Earth's continents concentrating the precious metals into ore deposits which can now be mined.

References

The tungsten isotopic composition of the Earth's mantle before the terminal bombardment. Matthias Willbold et al. Sept 2011.

Meteorite storm showered planet in gold. New Scientist Magazine. Michael Marshall. September 2011.

How to use Amino Acids to make Gold

In an August 2016 published study titled: Catalytic Application of Nano-Gold Catalysts Acids by Light Irradiation, that was written by Lilia Coronato Courrol and Ricardo Almeida de Matos, the authors created gold nanoparticles using only white light (*a xenon lamp*), Chloroauric acid, Milli-Q water (*creates ultrapure water*) and the amino acids valine, aspartic acid, threonine, tryptophan and arginine. As I shall go into detail later on in this book, key amino acids are valine and arginine.

What is also interesting is that when the Actinobacteria Rhodococcus is exposed to Chloroauric acid, it creates gold (Biosynthesis and Characterization of Gold Nanoparticles. Maria S. Kuyukina et al. Oct 2022). Actinobacteria are gram positive bacteria that help encode information in DNA. One of the functions of Actinobacteria is to help in the process of decomposition in organic substances. Another interesting thing is gold can be made from Saperavi red wine using chloroauric acid (Synthesis Of Gold Nanoparticles From Chloroauric Acid Using Red Wine. T. Pavliashvili Et Al. 2017).

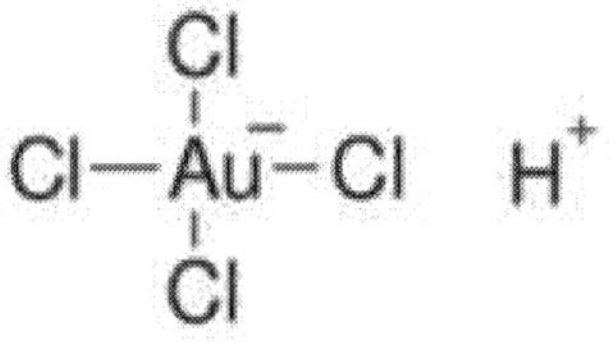

Chloroauric acid

Another research study published in October 1996 titled: An experimental study on gold solubility in amino acid solution and its geological significance, published by Zhang Jingrong and colleagues, stated that amino acids may have been responsible for the formation of source beds of gold. The researchers also stated that the proper concentrations of amino acids and temperature are key elements, with the pH being between 6 and 8 at 80°C giving the best results. Another paper, recently published in 2020 by Andrey A. Buglak and Alexei I. Kononov titled: Comparative study of gold and silver interactions with amino acids and nucleobases, stated the synthesis of silver occurs at an alkaline pH and that gold nanoparticles grow best at low temperatures in an acidic pH.

Later on you shall read how the Maillard reaction works and how it creates some of the most powerful anti-aging substances. The Maillard reaction involves the process of heat. An example is heating amino acids and sugars. This causes them to evaporate creating new or more concentrated substances. Getting back to the aforementioned paper (published in 2020), the authors of the study stated that gold nanostructure creation can occur via a reduction of amino acids in water. The best amino acids for this being: arginine, tryptophan, methionine, phenylalanine and threonine. The amino acid glycine is also used as a method to extract gold from minerals (Use of amino acids for gold dissolution. C.G. Perea May 2018) and amino acids are used to stabilize gold nanoclusters (Andrey A. Buglak. Sept 2020).

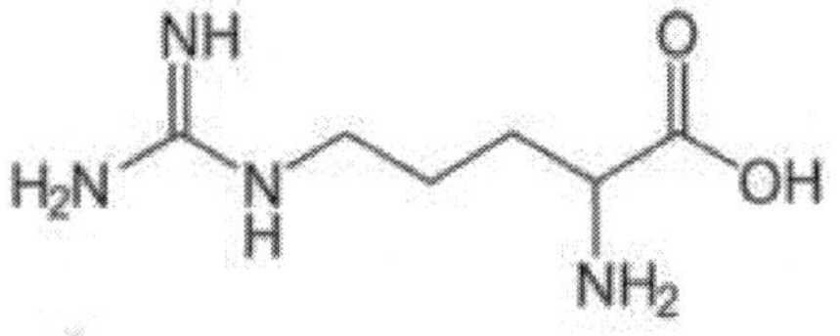

Arginine

Another study published in June 2023 titled: Amino Acids as Reducing and Capping Agents in Gold Nanoparticle, published by Aleksandra M. Figat and colleagues, observed the formation of a black precipitate while they were synthesizing the gold nanoparticles. If you are familiar with alchemy, the black precipitate is a key element in the making of the Philosopher's Stone. The authors found they were able to synthesize gold nanoparticles with the following amino acids: Glycine, Valine, Asparagine, Leucine, Alanine, Leucine, Phenylalanine, Serene, Threonine, Isoleucine and Aspartic Acid. Also protein cages and peptides have been used to synthesize gold. You can read more about this in the paper shown below under further reading.

Further Reading

Binding Preferences of Amino Acids for Gold Nanoparticles: A Molecular Simulation Study. Qing Shao and Carol K. Hall. July 2016.

Chapter 2
Alchemy and bioavailability

An ancient Zhuangzi text that was written between 476 and 221 BC contains anecdotes and stories which exemplify the carefree nature of the Taoist sage Master Zhuang (Zhuangzi). It is one of the founding texts on Taoism, along with the Tao Te Ching. In Chapter 1.3 of the Zhuangzi it states that on Guyi Mountain there exists a spirit-like man whose skin appears icy as snow, is as chaste as a virgin, and does not eat grains. However, he drinks the dew and sucks in the wind to maintain his physical immortality. It is his concentrated spirit which brings bountiful harvests year after year, saving things from corruption [65].

The Basics of Alchemy

Honolulu Community College (www.honolulu.hawaii.edu) Program 23. Lesson 4.1 on Alchemy states Alchemy is the cosmic art through which parts of the animal and mineral parts are liberated from their temporal existence in order to achieve a state of perfection. In the case of minerals; gold. For humans, immortality, redemption and longevity. These transformations are brought about by a material substance such as an elixir, the Philosopher's Stone by psychological enlightenment or by revelatory knowledge.

Gold has been made by bombarding mercury with neutrons, which was first accomplished by Sherr, Bainbridge and Anderson in 1941. This was repeated in 1980 by Glenn Seaborg who transmuted bismuth into gold. If you look at a Periodic Table of the Elements, bismuth is right next to the element lead. And today scientists are able to create silver nanoparticles using lead (Lead-germanate glasses: an easy growth process for silver nanoparticles. Ricardo Schneider et al. Aug 2017).

If you read the early literature on Alchemy, it states that in order to make the Philosopher's Stone, one would collect fresh dew from the grass by placing sheets on the grass and then wringing the dew out into a bucket. Next the dew would be put through a series of stages, many of which involve low heat. The time of year the collection of the dew was recommended was during the spring full moon and especially when the moon was in Taurus (refer to a Moon void of Course Calendar).

Research studies (66) have shown dew which settles on the grass contains significant levels of the amino acids glutamine/glutamate, proline and arginine. What is most interesting is the researchers in the study discovered that the amino acids showed a seasonal variation, with the amino acids in the dew peaking during March, showing above levels of proline and arginine. What is most interesting is proline has been shown to protect plants from heat (67).

Why the Bioavailability of nutrients decreases with old age

Bioavailability refers to the absorption, metabolism and activation of nutrients as they enter the body. The absorption of nutrients in the body fades as one grows older. Studies confirm that people of old age absorb less iron (68) and also exhibit reduced calcium absorption, which in turn affects their metabolism of vitamin D. (Barragry et al. 1978, Ebeling et al. 1992, Holick et al. 1989, Tsai et al. 1984). Also many elderly people are on prescription medication, which may decrease the way nutrients are absorbed (69). Also micronutrient and macronutrient (including protein) intakes drop significantly after age 65 (70). Studies also show that the bioavailblity of nutrients can be improved when the abundance of beneficial gut bacteria is at healthy levels or when there is a decrease of bad bacteria in the stomach (Bioavailability Based on the Gut Microbiota. Feng Zhang et al. Apr 2020). Also elderly people show a reduced ability to adapt to diets low in calcium (71) compared to younger people who are more able to rapidly adapt to such diets (Ireland and Dordtran 1973) and as shown earlier, the compound Carene, found in the 5,000 year old Bristlecone pine tree, has been proven to be extremely effective at keeping bones strong and healthy.

Studies also state (72) that some elderly people suffer from a malabsorption of healthy nutrients in foods. These include amino acids, carbohydrates, lipids, minerals and vitamins, which can lead to specific illnesses (72).

A lower nutrient bioavailability can cause a decreased capacity of the kidneys to convert vitamin D into an active form. A research study examined an aged Puerto Rican population near the city of Boston and discovered that only 18% of adults between the ages of 51 and 70 had healthy levels of Vitamin D (73). The bioavailability of certain nutrients is pH dependent. Nutrients that are not well absorbed due to a low acidic pH (74) include calcium, iron, beta-carotene, folic acid and vitamin B-12 (Russell 1986, Tang et al. 1996), (Camilo et al. 1996, Ribaya-Mercado et al. 1987).

Enhancing Calcium Bioavailability

Nuts are some of the very best anti-aging foods, especially nuts that are abundant in calcium. While many foods contain calcium, the key is eating nuts that have calcium that is easily absorbed into the body. The highest bioavailability of calcium from nuts comes from pistachios (75). Pistachios also have a high zinc bioavailability as do hazelnuts.

Chapter 3
Amino acids, Caloric Restriction & Lifespan

Amino Acid and Lifespan

Today people on the Greek island of Ikaria live 8 years longer than Americans, have half the rate of heart disease, experience 20% less cancer and experience almost no dementia (76) (77). The reason for their long healthy lives is because their diet is at least 90% plant-based and they obtain their protein from fish instead of meat. Beans, peas and chickpeas are also a staple part of their diet.

How Caloric Restriction lengthens lifespan

Tom Cruise is now over 61 years of age and looks very young for his age. This is because he follows a caloric restriction diet. In a 2021 Men's Health article titled: How Tom Cruise Remains Youthful at 61, and How You Can Too, which was published in July 2023 he stated in an interview that he uses a Beckham-devised diet. This diet consists of a total of only 1,200 calories, some grilled foods and almost no carbohydrates. Another person that has a calorie restricted diet is mathematician Courtney Brown (78). He is in his 70's and looks to be in his mid 40's. He talks about his caloric restricted diet in one of his videos, which goes into more details. Other people who look young for their age, Lenny Kravitz, Chuando Tan, Christy Turlington, Gwyneth Paltrow (known for her strict diet), Bianca Lawson, Gong Li, John Legend, Sandra Bullock, Pharrell Williams, William Shatner, Gwen Stefani and John Stamos.

I used to believe that skin gets dried out as one gets older. However this is not always the case. 90 year-old working supermodel Carmen Dell'Orefice, who has been married three times, still has tight firm skin and no sagging neck or jawline.

Research shows (79) that a significant increase in lifespan is observed when calories in the diet are restricted between 25% and

60%, compared to diets without caloric restriction (Borut Poljsak et al. Dec 2020) and a recent 2021 study (80) found that by reducing your calorie intake by up to 30%, it can considerably increase your life expectancy.

Average Calories per Food Group	
Food Group	**Calorie total per 100g**
Non-starchy veggies (beets, broccoli, okra and carrots)	15-50
Fruits (bananas, berries, apples and tomatoes)	16-85
Starchy veggies (squash, potatoes and corn)	19-99
Whole grains (oats, rice and quinoa)	68-122
Legumes (lentils, beans and peas)	118-170
Flour based foods (bagels, bread and pasta)	147-290
Dried fruit (dates, raisins and prunes)	237-298
Foods with sugar (maple syrup, table sugar, corn syrup and agave syrup)	259-410
Seeds & Nuts (cashews, flax seeds and walnuts)	473-660
Oils (coconut, canola and olive)	876-900
Plant-based nutrition for healthcare professionals: implementing diet as a primary modality in the prevention and treatment of chronic disease. Julieanna Hever and Raymond J Cronise. May 2017.	

How the right amino acids lengthen lifespan

Certain amino acids have been shown to decrease lifespan; with increasing concentration (81). These are phenylalanine, tryptophan, aspartate, tyrosine and valine. Three amino acids have been shown to exhibit the greatest increase in lifespan (at 5 mM doses). These were cysteine, asparagine and glutamine.

When cysteine-fructose is used in a Maillard reaction (82) it creates above average levels of antioxidant activity, as well as exerts metal chelating properties (the removal of toxic metals). We shall go into in-depth detail about the Maillard reaction later on, which also involves the chemical reaction taking place when proteins and sugars are both heated together.

Protein Toxicity

Nutrients which release acid precursors into your bloodstream are proteins and phosphorus, which includes the sulfur amino acid methionine. This can cause blood pH to decrease (become more acidic) (83).

Methionine restriction extends lifespan

Orentreich and colleagues tested their hypothesis that restricting the amino acid methionine in the diet could extend lifespan (84). Their study found that when mice were fed a methionine restricted diet that it resulted in increased lifespan of between 30% and 35%. The mice showed leaner bodies, improved glucose homeostasis, lower blood glucose and insulin levels and in studies on fruit flies (85), restricting methionine by 67% extended their maximum lifespan and mean lifespan by 2.4% and 10.5%. It was interesting that there exists a methionine restriction "sweet spot" as the researchers discovered that severely restricting methionine levels by 88% did not extend their maximum or mean lifespan any further. This sweet spot may vary between organism and species, depending upon various factors such as weight, lifestyle and other factors.

Methionine toxicity due to more acidic blood levels can be

decreased by adequate glycine intake (Luka et al., 2009). Indeed research studies now confirm that adequate glycine intake may reduce methionine levels. Diets supplemented with Glycine also exhibit anti-inflammatory and anticancer effects in studies on mice (86) (Alarcon-Aguilar, 2008; Wang et al., 2013; Zhong et al., 2003) as well as demonstrated beneficial effects in humans diagnosed with type II diabetes in a 3-month clinical trial (Cruz, 2008).

When 344 mice were given a Glycine supplement (87), they exhibited a significant extension in their lifespan (with glycine being given at levels of 8% (the best), 12% and 20% (Brind, 2011). Excess glycine caused weight loss.

Caloric restriction works because it reduces oxidative stress in the body, which in turn reduces free radicals occurring in mitochondria. This reduction substantially reduces proteins which become oxidized which contribute to mutated DNA (88).

Because excessive and incorrect proteins are one of the main causes of aging, caloric restriction is simply a way to reduce the amount of the wrong protein in your diet. Because most meats contain excessive levels of protein, which includes an over abundance of amino acids, a reduction of amino acids in the diet is one way to reduce the amount of calories in your diet. There are now some good studies that confirm that excessive red meat consumption can cause health concerns. For example a recent research paper published in October 2022 (89), which was a systematic review and meta-regression study, evaluated the relationships between health and the consumption of red meat. The authors in the study concluded there exists some evidence that the consuming of unprocessed red meat leads to an increased risk of mortality and disease incidence and a May 2019 study (90) reported that epidemiological studies show diets high in animal proteins, particularly red meat, contains high amounts of methionine and that this may be related to age-related diseases. William Shatner writes in his 2018 book titled: Live Long and.... that he has significantly cut back on heavy meat consumption. Perhaps this is why he is one of the world's oldest living actors and doesn't look 92 years of age.

When researchers examined the diet of one of the longest lived populations, the Japanese Okinawans, they discovered that they consumed fewer total calories. This was because they ate significantly less barley, wheat and other grains [(91)].

When we look at what exactly is in meat that is causing the health problems, we see that it is due to the protein amino acids isoleucine and valine, possibly due to them being subjected to high temperatures [(92)].

Out of all the common household oils studied for producing toxic compounds during the Maillard reaction (sunflower oil, extra virgin olive oil and canola oil), cooking in sunflower oil produced the least number of heat toxins [(93)]. Cooking in the sunflower oil created more Methylpyrazine, which is found in peanuts, coffee and red peppers. The cooking in the sunflower oil also produced more dimethylpyrazine, then the other oils, which is found in coffee.

So what happens if we cook some commonly eaten foods in sunflower oil? Would it show an increase in their nutrients? Studies by Hedrén et al. found that when sunflower oil was used to cook certain foods that they all showed increases in their beta carotene levels. Increases were observed in pumpkin (64%), cassava (47%), sweet potato (39%), boiled pumpkin (19%) and **boiled carrots (74%)** [(94)]. These are significant increases and it means the less toxins that are created during heat, the better the food is for the body.

A longitudinal study involving 2,983 men and women between the ages of 40 and 65 had their plasma levels of lycopene, carotene, lutein, zeaxanthin, and cryptoxanthin measured. The researchers found that people who had higher levels of these carotenoids showed higher cognitive scores. This study shows that people who have a diet with healthy levels of carotenoid-rich foods show stronger cognitive functioning as they grow older (*Lycopene and cognitive function. Kristi M. Crowe-White et al. May 2019*). As a side note, mango is not only rich in carotenoids, but also carene

(*Chemical Composition of Mango. Maria Elena Maldonado-Celis et al. Oct 2019*).

What is also interesting is when you do Oil Pulling, it has been found to have numerous health benefits. Oil Pulling is when you swish your mouth with sunflower oil for a period of time until it turns white then spit it out and rinse out the mouth. Oil Pulling is the main antidote for enjoying good teeth and gum health, and in some cases to help alleviate toothache, which I go into more detail in my book titled: The Complete Guide to Natural Toothache Remedies and Re-mineralization.

And as we shall explore in an upcoming chapter, when Ginseng is heated, the amino acid valine was one of the amino acids that was easily impacted by the Maillard reaction (95) and Valine has been shown to extend the lifespan of mice (96). This means that plant protein amino acids in plants undergoing chemical reactions involving the proper heat, create mostly beneficial reactions. However, in the case of meat products, when meat is cooked, especially at high temperatures, it not only forms AGE's, but the valine may be creating more negative substances that contribute to ill health. Hence, there exists a "sweet spot" for the substance being heated where the amino acids undergo their phase transition to become beneficial substances for the human body. Let's examine the data to gather strong evidence to support this hypothesis.

The Branched Chain Amino Acids

Isoleucine and Valine are in the class of amino acids called Branched Chain Amino Acids (BCAA's). There now exist numerous scientific research studies involving humans, flies and rodents (97) that prove that reducing these amino acids in the diet extends lifespan.

So if we want to go one step further and eliminate the main proteins that are responsible for accelerating aging, we can look at the scientific research. The research shows (98) that when the amino acids isoleucine and valine are removed from the diet, it extends lifespan. The research also shows (99) that the amino acid leucine is problematic, however this protein amino acid does not appear to be responsible for accelerating the aging process.

Foods with extremely high levels of Leucine include (100)
Adult bovine's rump, baked ham, bresaola, chicken breast, deer, guinea fowl, pork, rabbit, asiago, grana cheese, parmesan cheese, drained tuna, mullet roe, smoked salmon, raw dried broad beans, arachid butter, pine nuts and unsweetened cocoa powder.

Besides valine, leucine is also a branched-chain amino acid. Studies on both mice and humans (101) have discovered that diets low in protein are highly beneficial. A research study found that a diet that had low isoleucine levels was beneficial for the liver, restored homeostasis functioning and resulted in more energy in the body through the production of ketone bodies. There were other data points in the study that suggested that the reduction of these two amino acids greatly enhanced physical energy in the body. Hence, if you are vegetarian and want to have more energy, than a reduction in these two amino acids is key. The authors in the study concluded that reducing isoleucine be recommended as an approach for treating and preventing diabetes and obesity.

How to Increase your energy level if you are Vegetarian
From my own journey of becoming vegetarian, the very first problem I encountered was a lack of physical energy. This subsided after a few months because my body eventually adjusted and my energy returned back to normal levels. However when my energy returned, it returned stronger and more clearer than before. If you are just starting out vegetarian, the chart on the following page shows some non-meat foods that will help get you through the temporary energy crisis. If you want to learn more about becoming vegetarian, my book titled: The Vegetarian's Guide to Longevity via Gene Therapy and Raw Foods has all the information you could possibly ever need regarding a healthy vegetarian lifestyle.

How to get proper protein levels -
Skim Milk & Tuna contain adequate protein if you are vegetarian or vegan (102) and don't mind eating fish (pescatarian).

Protein Levels in plant based foods		
Wheat 32.3%	Brown rice 20.6%	Corn 7.3%
Potatoes 3.1%	Soy (includes Tofu) 2.7%	Peas 1.0%
Oatmeal 0.3%		
Protein content and amino acid composition of commercially available plant-based protein. Stefan H. M. Gorissen et al. Aug 2018		

Additional substances that contain an abundance of plant protein are calcium caseinate (38%), casein (34%) and eggs (32%) (103).

Some people may think taking the amino acid supplement Whey Protein will help give them enough protein, however research shows that Whey Protein contains high levels of isoleucine, leucine and valine (104) and studies on mice show that a reduction of isoleucine in the diet increases lifespan (105).

Vegetarians naturally have higher concentrations of glycine. Researchers speculate that the consumption of meat may reduce levels of glycine in the body (106). This was shown in a clinical trial (107) where glycine levels declined slightly after eating meat for a week, even when glycine levels in the meat were 50% higher, compared to glycine levels in the vegetarian diet. Also another study found that a diet which included 8% glycine (109) led to a statistically significant increase in lifespan in both female and male mice. What was surprising in the study was that the mice were less likely to die of lung cancer, suggesting that glycine exerts its protective effects via lung homeostasis.

In summary, the best high physical energy foods (from highest to lowest) are - Microalgae, Soy, Brown Rice, Oats, Lupin, Wheat and Corn.

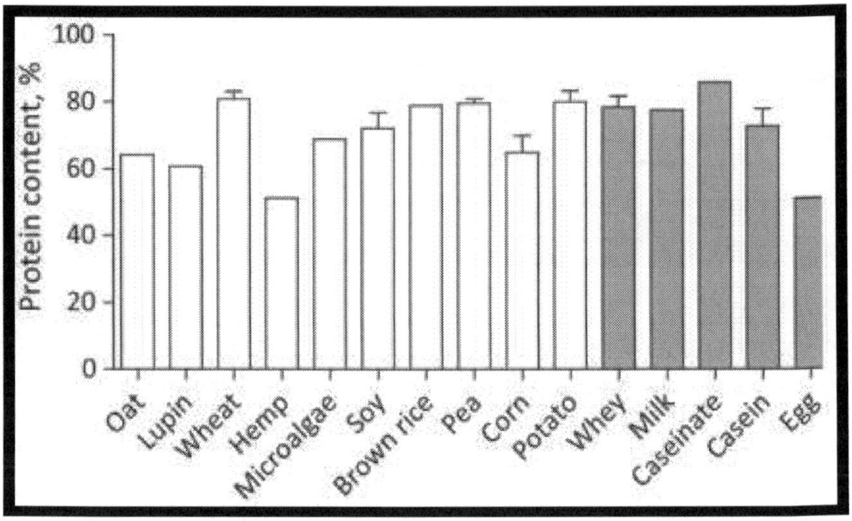

Above graphic courtesy of: Protein content and amino acid composition of commercially available plant-based protein isolates. Stefan H. M. Gorissen, et al. Aug 2018.

So what does the data say regarding how much amino acids in the diet should be reduced? Research has found (109) that when flies had a 50% reduction of branched chain amino acids in their diet, that they exhibited an increase in their lifespan. When these levels were increased to 85% their lifespan also increased, but they were more prone to sterility (*a reduced capacity to have offspring*). In a recent research study (110) published in July 2021, the optimal reduction of branched chain amino acids in mice was 67%. This gave them an increased lifespan of up to 30%. When it comes to humans, the percentage of reduction of BCAA's is similar, being 40% (111).

The Amino Acid Methionine. Victor or Villain?

Another amino acid which can be detrimental to the body in large amounts is Methionine. Natural energy foods that contain low amounts of Methionine include microalgae (0.0%), oatmeal (0.2%), lupin (0.3%), peas (0.4%), soy (0.4%) and wheat (0.9%).

When glycine is added to diets containing an above average

level of methionine, it has been shown to suppress methionine levels [112]. When the amino acid methionine enters your body it is than converted into homocysteine which is a substance that has been associated with accelerated aging and aging-related disorders. The amino acid that comes to the rescue is glycine, which reduces methionine levels in the blood [113]. When the amino acid methionine was reduced by between 30% and 40%, it extended lifespan in studies on mice [114] and when glycine was given to mice, it extended their life because it suppressed their levels of methionine [115]. This may be why diets supplemented with glycine exhibit anticancer and anti-inflammatory effects in mice. (Alarcon-Aguilar, 2008; Wang et al., 2013; Zhong et al., 2003) as well as be beneficial to patients diagnosed with type II diabetes. Significant extension of lifespan was also observed when mice were administered glycine at 8%, 12% and 20% in their diet (Brind, 2011) [116]. When the mice were given above average amounts of glycine, they lost weight. The glycine "sweet spot" that obtained the longest lifespan was ($p = 0.03$) glycine supplementation at 8%. Glycine has also been used to heal skin wounds, reduce alcohol overdose, treat leg ulcers, treat ischemic stroke and reduce the harmful effects of kidney drugs after an organ transplant [117].

Another study [118] hypothesized that glycine supplementation worked to reduce the levels of oxidative stress created by methionine. When the glycine was administered to mice at 8% or 12%, it increased their median lifespan from 88 weeks to 113 weeks; with their maximum lifespan increasing from 91 weeks to 119 weeks (45 months). The average lifespan of a mouse is 24 months. This is a significant finding, proving that the restriction / neutralization of problem amino acid proteins extends lifespan.

Glycine Synergy

When Glycine is combined with lysine and arginine and given as an oral supplement, it has been shown to lower cholesterol levels up to 51% [119]. The best ratios were when the amino acids Arginine/Lysine and Glycine/Methionine ratios were at the ratios of 1:1 or 2:1.

Foods that meet this ideal amino acid ratio (120) include: whole dried egg, sweet potatoes, oysters, kale, spinach, bananas, asparagus, lotus roots, mung beans, oatmeal, corn, beets, egg yolk, beet greens, broccoli, dates and oranges.

Foods abundant in Glycine

Foods that contain an abundance of Glycine (121) include: grain products and milk and dairy products such as yogurt, kefir and chicken. Other foods include wheat, with the highest levels of glycine being in Hungarian wheat and the lowest in UK grains (122) and rice has been found to contain about twice as much glycine compared to proteins found in vegetables or animals (123). Regarding spices, caraway seeds contain extremely high levels of glycine and lysine (124). Caraway seeds were found to greatly enhance the ALDH2 gene (125), which is the gene that protects your body's organs against oxidative stress. Which may be why caraway seeds are used to treat chronic fatigue syndrome (126).

Chapter 4
Exploring the Maillard reaction and its relationship to Alchemy

During my research over the decades I have discovered that the most successful anti-aging substances only require a small amount to be ingested and they do not spoil. This is what the red powder of the Philosopher's Stone was all about. It is a red powder with a glassy shiny reddish hue which was easily transported, never spoiled and when only a few sprinkles were added to wine, would grant the drinker physical immortality. A book I highly recommend if you are interested in pursuing alchemy is the book titled: The Book of Aquarius eBook by anonymous. This book contains snippets of all the best parts of books written on Alchemy throughout the last 200 years or so. What is most interesting is from my own experience, when I make an anti-aging extract when the moon is in Aquarius (consult a Moon Void of Course Calendar for when the moon is in Aquarius), that the potency of the extract lasts much longer. This is just from my own experience and it would be interesting to see some research studies that would confirm this. If you want to learn the art of extract making, I highly recommend my book titled: The Official Guidebook of How to Make Tinctures and Alchemy Spagyric Formulas, on how to make your own extracts.

The fundamentals of a successful anti-aging substance should consist of the following -

1 - Be easily transportable

2 - Not spoil easily or keep for long periods of time
3 - Easy to make from easily accessible substances

4 - Only a small amount needs to be consumed in order for it to exert its anti-aging properties

Wine fits many of the above criteria, as do some quality supplements and extracts. Many of the best substances that exert anti-aging effects are dose dependant or have a sweet spot for maximum effectiveness. This simply means taking too little does no good, or taking too much causes harm. For example if a person added too much of the red powder to wine, their hair, nails and teeth would fall out, eventually growing back. Think of this as exercise, where just the right amount of exercise is good for you, but too little or too much can be unhealthy. You can find out the sweet spot of an anti-aging substance by carefully feeling how rejuvenated your body feels the following morning after taking an anti-aging substance before bed, how fast you recover from heavy exercise or by how much energy you feel in your body a few hours after taking it.

The Maillard reaction and the Philosopher's Stone

Whether or not you believe in the story of the Philosopher's Stone, there are some very interesting correlations between substances that when they are heat-treated have significant effects on lifespan; many showing significant increases in antioxidants. This process involves what's known as the Maillard reaction. During food processing and storage, the Maillard reaction has a major impact on the color, flavor and nutritional value of products. It forms desirable candy like flavors, a golden caramel color, strong baked aromas as well as generates and increases antioxidant levels [127].

The Maillard reaction is simply a chemical reaction that occurs when amino acids are heated with a sugar. An example is sucrose, which is than heated to a specific temperature. The key is to keep the temperature at a certain level, or else the substance can than turn into toxic AGE's, which actually contribute to the aging process.

Maillard reactions and Transmutation

Let's first explore some simple Maillard reactions that you encounter every day. The palladium in your car's catalytic converter converts harmful exhaust to a harmless substance by the reduction of nitrogen oxides. A similar process takes place when sugars (*sucrose*) and amino acids are combined. When both are heated together, much of their existence evaporates, producing multitudes of new compounds. Roasting coffee beans is one example. When Lotfy et al. heated Quinoa protein at various pH levels from 5 to 9 (129) he noticed changes in the sensory attributes, going from burnt-coffee to caramel.

The Maillard reaction was coined in 1912 by the French chemist Louis Camille Maillard (128) who was studying the reaction taking place between glycine and glucose. Research studies have shown that when ginseng is heat-processed with certain amino acids, it produces an abundance of the ginsenosides Rg3, Rg5 and Rk1 (Effect of Amino Acids on the Generation of Ginsenoside. Jun Yeon Park et al. Mar 2016). Ginsenosides are the most active pharmacological components in ginseng. These three simple ginsenosides have been scientifically proven to fight lung cancer, reduce inflammation, improve breathing and promote the proliferation of human endothelial cells. This simply means it creates better health of cells which line the inside surfaces of body cavities, lymph vessels and blood vessels. The authors of the research study concluded that the heat treated ginseng exhibited increased medicinal effects. This research experiment is covered in more detail in a Chapter 6, which also includes full cited references and how they did it. When we dig a little deeper into the research literature, we find that Rg3 is also found in Centella asiatica (Gotu Kola) (130), which also happens to contain the substance Rg5 (131).

If we want to find out what happens to Gotu Kola when it is heated, we discover that it generates a **whopping 200% increase in polyphenols** compared to Gotu Kola leaves that are unheated (132).

A look at the longest living people and their diet

These two people we shall look at were reported in Time Magazine to have lived to over 150 years of age. In the future, x-ray machines that can penetrate the soil and look at human DNA using non-invasive methods may be able to verify these claims, but let's look at their diets and see if we find any interesting correlations between foods in their diet that were heated and lifespan.

When we research one of the world's longest living people, Mr. Li Ching-Yuen, purported to have lived to over 256 years of age, the literature shows that Gotu Kola was one of the main herbs in his diet (133). Besides Gotu Kola, he is reputed to have consumed rice wine and goji berries (134). A few months before he died, he was quoted as saying in Time Magazine (135) "*the key to long life is to keep a quiet heart, walk sprightly like a pigeon, sleep like a dog and sit like a tortoise.*" Could the reason Mr. Li Ching-Yuen lived so long was because he heat treated the Gotu Kola? This gives a strong confirmation that the making of the red powder of the Philosopher's Stone involved a special process involving heating the right type of substances. If we examine another person, Mr. Zaro Aga, who was reported to have lived to 170 years of age and who worked as a coffee brewer for the local city council in his local town in the country of Turkey, (136) we observe another correlation between the heating of a specific substance and longevity. When coffee beans are roasted, they exhibit an increase in their anti-aging substances (137). This is because coffee beans contain certain amino acids, that when heated to the right temperature, create the Maillard reaction. When the amino acid lysine-glucose is heated for 60 minutes, it will show increased levels of antioxidants (138) and when black cumin seeds (*which are a native plant of Turkey*) are roasted, they yield higher oil quality (139). This increase in antioxidants has also been observed in coffee beans that are roasted. This increase is dependent upon the beans' temperature and roasting duration (51) (140). In other words, there exists a "sweet spot" for creating anti-aging substances in coffee via the Maillard reaction.

Research using various models of aging has found that lifespan can be prolonged by some antioxidants (141). This could be why numerous studies show coffee extends lifespan with some studies showing up to a 52% increase in lifespan (142) (143) (144).

Recommended Reading
Confirmation that the Maillard reaction is the principle contributor to the antioxidant capacity of coffee brews. Liu Y., Kitts D.D. Food Res. Int. 2011;44:2418–2424. doi: 10.1016/j.foodres.2012.037.

Antioxidant Levels in Everyday Foods

Name	Antioxidant Level
Oranges	0.9
Papaya	0.6
Dry Crushed Leaves Of The Baobab Tree	48
Dried Plum	3.2
Fresh Dog Rose	24
Dried Dog Rose	**78**
Raw Apple	0.4
Dried Apple	**3.8**
Pomegranate	1.8
Baobab Tree Fruit	10.8
Prunes	2.4
Dried Apricot	3.1
Fresh Leaves Of Cabbage Tree- Stem (Moringa Stenopetala)	3.7
Dried Leaves Of Cabbage-Tree- Stem (Moringa Stenopetala)	**11.9**
Raw Kale	2.8
Cooked Broccoli	0.5
Dried Mango	1.7

Strawberries	2.1
Red Or Green Chili	2.4
Dried Dates	1.7
Black Olives	1.7

The total antioxidant content of more than 3100 foods, beverages, spices, herbs and supplements used worldwide. Monica H Carlsen et al. Jan 2010.

Phloridzin and the Maillard reaction

There exists an extremely large amount of research showing the anti-aging effects of polyphenols (145) and I cover this in greater detail in my other anti-aging books. For example, the bark of the apple tree contains an abundance of Phloridzin, with the Spartan variety of apple having the highest concentration of Phloridzin compared to other varieties (146). I shall cover later on why bark is such a powerful anti-aging substance, mainly due to the fact that it has to create substances that protect the tree from insect attack as well as absorb life giving nutrients from the air. Extracts of apple bark have been shown to contain an abundance of phenolic polyphenols.

Phloridzin is part of the polyphenol family which includes the substances quercetin, catechin, epicatechin, procyanidin and rutin, all of which are powerful anti-aging substances. Phloridzin levels are abundant in dried apples and apple juice (147). Dried Mexican oregano contains 136 mg per 100 grams of Phloridzin and it is found in many longevity foods such as the pomegranate and prunes (147). It is also found in cranberries, cranberry products, rosemary, thyme, oregano, sage, basil and hazelnuts. What's interesting is that Phloridzin undergoes a radical transformation during the Maillard reaction. If Phloridzin becomes exposed to heat, it exhibits significant increases in its antioxidant levels (148). This is why Phloridzin has been classed as a potent Anti-Glycation Agent (149). An Anti-Glycation Agent (AGE) is any substance that protects the skin from damage and aging. You can also reduce aging by reducing the AGE's in some foods. Phloridzin obtained from apples has been shown to kill cancer cells. (150).

Heat Treated Sesame Seeds

Sesame seeds contain Sesamin, which is a powerful anti-aging substance with many studies confirming its effects on extending lifespan (151) (152) (153). When Sesame seeds are roasted to the right temperature and for the right amount of time, they will increase their phenols (a class of polyphenols) up to 32.4% (154).

The Healing Power of Asparagine

Asparagine supplements have been shown to improve nonverbal communication and attention span (155). If you see Goji berries in your local co-op or organic food store you will see they are dried, but with a resinous coating. This long term storage is what contributes to its abundance of Asparagine in Goji berries (156), with the highest levels found in the SL cultivar variety, especially when the berries are harvested during October (156).

Asparagine does not appear right away in fresh fruits. It appears as a by-product of fermentation of acids in the fruit. After a few months of storage, asparagine will begin showing up in the freshly stored fruit. Asparagine is found in abundance in asparagus that has been heat dried followed by dried apricots (including heat dried or standard dried apricots), heat dried apples, heat dried bananas, fried red potatoes, red pepper and bell pepper (157). Interestingly if the asparagus is boiled or ate raw, it contains some of the lowest levels of Asparagine. Therefore a good combo would be boiled asparagus with dried apricots or apricot sauce with goji berries.

Asparagine does not produce other amino acids, but instead acts as a protein synthesis moderator. If mammalian cells become deprived of glutamine, the amino acid asparagine will do its best to keep the cells alive. A study hypothesized this effect is due to the asparagine enhancing protein synthesis (158).

The Secret to Mr. Run Run Shaw's Longevity

In Roger Moore's 2014 autobiography titled: Last Man Standing. Tales from Tinseltown, he recounts the meeting of Mr. Run Run Shaw, who was in his 30's. Mr Shaw founded a film company that filmed Kung Fu movies. Roger Moore stated that the secret to Mr. Shaw's health and longevity was from taking 3 ginseng pills every morning after getting up and another 3 before bed. He would take the ginseng pills with Scotch whisky. Mr. Shaw stated to Roger that this allowed him to do everything at 70 years of age, the same as if he was 30 years of age. Mr. Shaw stated that the Ginseng came from mountains in mainland China where it would be dug out with small spoons when harvested so that the root was preserved intact when it was harvested. Mr. Shaw stated that the powder was ground up into a powder and made into capsules before Mr. Shaw consumed it. This is because grinding up the entire root and making it into a powder all at once would cause the special rejuvenation nutrients in the ginseng to be lost. It is very likely that the ginseng root was made into powder every few days directly from the harveste d root and then taken fresh, leaving the rest of the root intact to make more powder from. This is because Mr. Shaw had showed Roger a ginseng root that he kept in a special wooden box. Mr. Shaw passed away in January 2014 at 106 years of age.

Chapter 5
Exploring the lives of past Immortals

As stated at the start of this book, during the biblical era, people regularly lived hundreds of years, and as stated previously, when Greg Bradden interviewed some of the oldest people living today, he stated it was because they practiced having a "calm heart". A calm heart would lead one to lead a life that was more pure and free of toxins produced by out of control emotions. If we examine the anti-aging texts of the past, we see many references to "purification".

I strongly believe from my decades of research that this purification is an inner form of purification that comes from practicing what's known as Internal Alchemy. From my years of learning how to become a master at practicing the art of Internal Alchemy, from the concepts outlined in my book titled: Emerald Tablet Number XIII. The Keys of Life and Death. The Energy of QI Gong, Microcosmic Orbit and Hands on Healing, I am strongly convinced that it not only neutralizes negative emotions (purifying the inner part of one's being), but also positively affects the environment and surroundings that one practices it. This creates new fields of healing energy which positively affect plant growth, calmer local weather, reductions in crime and it contributes to an overall environment of harmony and well-being.

The Upanishads write that a physically immortal state is obtained via purification and a sublimation of the 5 elements (earth, fire, air, water, ether) which sustain the body. In Chapter 2, Verse 12 of the Shvetashvatara Upanishad, the text says a yogi's body is purified by the fire of yoga, which allows him to be free from old age, illness and death. Only when earth, fire, air, water and ether are able to be made manifest by the yogi, can this take place. An immortal was said to be able to instantly heal wounds, pacify animals, extinguish large fires and turn the course of rivers by his mere breath alone (Campany,2002, p. 21). If you read the book of Aquarius by anonymous, which is one of the best books on how to make the Philosopher's Stone, it is interesting to note that one of the substances used to make the Philosopher's Stone was dew. This was done by collecting early morning dew from the grass during the start of spring. I hypothesize that the ancients were able to purge their body of unclean negative spiritual parasites through the process of mastering Internal Alchemy. After having mastered the art of Internal Alchemy, they then went onto creating the next stage of Alchemy, the making of the red powder; the Philosopher's Stone. This means the making of the red powder is an act of purification, of purifying specific substances, with part of the procedure involving low heat, which than transforms the substance into a bioavailable agent allowing the body to absorb the life-giving nutrients in red wine.

In the text titled: Immortality in the Daoist Tradition (159), which was written by Pablo Vazquez in 2019, it states the human body contains within it multitudes of spirits and interloping monsters which contribute towards a person's progress and physiological makeup (Wong, 1997, p. 54). If the human body cannot maintain harmony with the good spirits in the body, it slowly creates negative emotions, which in turn cause adverse health conditions or even death (Wong, 1997, p. 54). The text goes on to state that these toxic beings can be purged from the human body through specific physical regimens, rituals, spiritual activities, or in certain extreme cases, exorcisms.

Chapter 10 of the Dao de Jing refers to how breathing can be used to reclaim this harmony. "*In concentrating one's breath, are you able to become as supple as a baby? Are you able to polish your mind, and leave no blemish?*" Other techniques (160) which expunge these undesirables include specific breathing techniques, meditation, special plant based diets, interior alchemy or adhering to a just and moral lifestyle (Philtar, 2019).

Angelica Dahurican root

This plant was used by the famous military physician Zhang Cong Zhen (1156 to 1228). He believed disease was the result of evil factors (pathogens) which unknowingly entered the human body (*Wikipedia. Angelica dahurica*). Zhang Cong Zhen recommended Angelica dahurica (Bai Zhi) be used to purge the body of negative influences such as dryness, cold on the skin, heat or clamminess.

What is most interesting is fermented Angelica sinensis has been shown to increase the ratios of Firmicutes/Bacteroidetes in the body by 95% (*good bacteria*), enhance NRF2 and restore healthy levels of microbiota in the intestine (*good bacteria*) which decline due to the effects of aging (Fermented Angelica sinensis activates Nrf2. Xuerui Lu et al. Jan 2023). An example of a lifestyle that adheres to a lifestyle of balance and a calm heart involve 3 simple principals.

Angelica dahurica

1 - **ENVIRONMENT**. The person identifies their specific situation that exists in their environment, including their physical makeup and geographical situation. They would then perform the proper exercises related to their environment and ingest foods specifically designed to adjust to the natural rhythms of the yin and yang as they occur throughout their surroundings. There are always anti-aging plants that are growing within at least 30 miles of where one lives that can be used to heal and exhibit anti-aging traits. This allows one to be aware of any diseases or disorders that may develop, allowing them to conserve their vital energy. For example, when I wrote my book titled: Medicinal and Edible Healing Plants of Australia's Sunshine Coast and the Noosa Heads Wilderness, I was surprised to learn of the many anti-aging plants that grew nearby where I lived.

2 - **SPIRITUAL NOURISHMENT.** Subtle forms of cosmic nourishment are necessary to keep the body free of negative energy which causes illness. One method is by eating the proper food which allows one to align themselves to the rhythms of the cosmos. This in turn allows one's body to absorb the life giving energies of the moon, sun and specific constellations which emit powerful rejuvenating energy at specific times of the month (see my book titled: adhering). This alignment of oneself with the underlying cosmic rhythms of nature allows one to experience higher states of consciousness which is necessary for adhering to the path of longevity.

3 - **DIET**. Attention, motivation and discipline is paid to diet and food intake. This creates a strong link to nonhuman cosmic agencies, which include the nature gods. However one refuses to become part of it, knowing man is allowed to exert a certain amount of control over nature. Their bodies become residences and pathways of the divine, as their spirits merge with the larger universe as a whole. Hence their place of residence becomes the sky with its rotating planets and celestial palaces. At this stage ,the immortal no longer needs to languish in the underworld, but regularly ascends to meet his or her celestial partners.

When we examine the Chinese texts during the time of the Chinese Dynasties, we come across men who were said to be physically immortal and that lived in the mountains. In principle, a remote mountain region would be a place where one would be more able to keep a calm heart, compared to the noise and bustle that exists in a large city. So let's explore some of the Chinese immortals of the past.

Many of the people I am about to mention were part of the group of the 8 immortals. Each power of an immortal can be transferred to a vessel that destroys evil or it can create new life. Many were born during the Tang (618 to 907) or Song Dynasty (960 to 1279). The scholar Ge Hong who lived during the Jin Dynasty (266–420), stated Mount Ziluo in China exists as one of the main supernatural domains of physical immortals.

Zhang San-Feng who was born in Shaowu, Nanping, Fujian during the end of the Southern Song dynasty (1270 C.E.) was said to have a lifespan of 211 years (161). Zhang obtained the knowledge of immortality at age 67 when he met Taoist Huolong Zhen Ren in the Zhongnan Mountains (162). He was then taught how to practice Internal Alchemy. Today if you visit the Jintai Temple in China (*pictured*) located in the Baoji Mountains, you can see displays of legends of Zhang in the temple.

The long lived person known as Pengzu is comparable to Methuselah in the West. He is well known for his herbal treatments. Legend says he married over 100 times and fathered hundreds of children when he was well into his 450th year (163).

Zhang Guo Lao was a mountain hermit said to have lived for hundreds of years. He was known for drinking wine and good living, as well as being a protector of children (164). Zhang Guo Lao was also said to be capable of performing resurrections and transmuting metals (Reninger, 2018).

Lü Dongbin had retired after working for the government. While at a wine store he met Zhongli Quan (165) who taught him how to make the Golden pill, whereupon he lived to two hundred and twenty years (165) (166).

The Confucian scholar Han Xiang Zi was a patron of musicians and one of the 8 Chinese immortals (167) (168).

In the book titled: The Story of Han Xiangzi. The Alchemical Adventures of a Daoist Immortal by Yang Erzeng, Master Lü says "*Everyone can be as long-lived as Heaven and Earth. Only people are not aware of it*." The book goes on to state that the mind of humans is constantly restless, and asks the question, "*Can humans have a constant mind that holds fast to the Dao?*" meaning that a calm mind creates a calm heart which creates a life free of excessive toxic emotions.

Many of the early teachers of Internal Alchemy could only teach students who had the mental discipline to adhere to specific diets. This included no eating of grains, pork, scallions, leeks, raw fish and dog. Many teachers stated that the eating of grains was strictly warned against because it could be the cause of old age and most illness (Campany, 2002, pp. 22-23).

While there are many conflicting studies regarding bad health being associated with rises in glucose levels and grain consumption, one study (169) hypothesizes that the rise in insulin levels may be attributed to long term consumption of refined grains as well as red meat. The authors in the study found that a diet high in refined grains and red meats caused significantly higher levels in the participant's insulin secretion levels, compared to those who ate the diet containing chicken, nuts, dairy and wholegrain. Studies on mice (170) confirm that reducing insulin levels by 25% extends lifespan.

Further Reading

A Whole-Grain Diet Increases Glucose-Stimulated Insulin Secretion Independent of Gut Hormones in Adults at Risk for Type 2 Diabetes. Steven K. Malin et al. Mar 2019.

Chapter 6
Merging Alchemy with Modern Day Longevity Science

Everything is composed of moving and circulating fields of energy which throughout its very existence is always being transformed into new types of matter. It cannot be destroyed. Heat is one example that drives changes through chemical reactions. In this chapter, we shall start to dive a little deeper into the mystery of how heat creates a chemical reaction, which in turn produces a substance that can create substances to help offset the effects of aging. This process is very similar to that of creating an antidote. In the case of snake anti-venom, a small amount of snake poison is injected into a live sheep or horse. After a period of time the sheep or horse begins producing anti-bodies as a biochemical reaction of the animals' immune system fighting the poison. These are called anti-bodies which in turn are harvested from the animal and used as a snake bite antidote [171]. Let's first begin looking at the research study of treating ginseng with heat. Ginseng is a plant that produces red berries and is world renowned for its anti-aging properties.

When Korean Red Ginseng is fermented, which is really similar to making a simple water extract or tea, it has been shown to lower Insulin levels in old-aged mice [172]. Another positive benefit of fermented ginseng is it increases the levels of the ginsenoside metabolites Rk1, Rg3, Rg5, compounds K, Rg2, Rh1 and F2 and its flavonoids [173].

When something is fermented, it creates an exothermic reaction, which causes it to give off its own form of internal heat. So let's follow in the footsteps of the old Alchemists and explore further to see what happens when we heat Ginseng with amino acids.

A research team [174] obtained 4 year old dried Ginseng from a local market and ground it up into a powder until it was able to pass through an 80-mesh sieve. Next it was made into an extract and then filtered through filter paper (Advantec, Tokyo, Japan), than evaporated to create the dried ginseng powder. During the heating phase, the amino acids glutamine and leucine were added to the Ginseng powder to create the Maillard reaction. The heat processing method allowed ginsenosides in the ginseng to react with amino acids. The combination was than steamed at 120°C for 3 hours than dried at 50°C for 72 hours.

RG3

After the final combination was analyzed, the researchers found that the ginsenoside Rg3 in the ginseng had increased to between 50 and 100%. When the Ginseng was than heated with the amino acids valine, arginine, and alanine, the levels of ginsenoside Rg3 also increased compared to the controls. Valine has been shown to extend lifespan in mice (175). So these substances that became amplified via the Maillard reaction via the heat processing must have anti-aging effects. Let's look at the data to confirm this.

Small amounts of Rg3 have been found to help maintain the cells which line the inside surfaces of body (176). This includes blood vessels, cavities and lymph vessels. These are all parts of the human body that are related to healthy circulation. So this particular Maillard reaction of Ginseng creates a substance that promotes healthy circulation in the human body. Rg3 is also being studied for treating fatigue without using harsh, expensive or addictive pharmaceuticals (177). This makes sense because any substance that enhances circulation in the human body would also allow a person to experience more physical energy. Rg3 is also scientifically proven to protect the human body against the dangers of oxidative stress (178). What is most interesting is Ginsenoside Rg3 has been shown to reduce inflammation in ganglion cells in the eyes (179). Inflammation and oxidative stress are a part of the aging process and can contribute to numerous chronic disorders and degenerative conditions. These include arthritis, neurodegenerative and autoimmune disorders, inflammation, cancer and cardiovascular diseases (180). Rg3 has also been scientifically proven to protect the body against DNA damage (181).

RK1 and RK5

Heat treatment of Ginseng with the amino acids glutamine or leucine has been shown to result in an increase of the ginsenosides Rk1 and Rg5 (182) (183). Rg5 reduces breathing congestion, reduces inflammation and reduces ROS (184). ROS is a term for Reactive Oxygen Species, which is a major contributor to free radicals. Rk1 and RK5 both have lung cancer stem cell fighting properties (185) (Hyunhee Kim et al. Jan 2021).

RG1

Ginsenoside Rg1 enhances the expression of the SOX9 gene (186) and reduces intraocular pressure of the eye (187). The latest anti-aging literature is showing that when the SOX9 gene is stimulated, it starts a series of molecular anti-aging processes throughout the human body (188). For example, elderly individuals exhibit reduced levels of SOX2 (188) and SOX9 levels are lower in mammals who have reduced cartilage in their bodies (188). Ginsenoside Rg1 has been shown to stimulate genes related to collagen production as well as the SOX9 gene, reducing inflammation up to 50% (189).

RB1

The ginsenoside Rb1 is present in abundance in ginseng. Out of all substances that protect the body from osteoarthritis and promote cartilage growth, RB1 has been shown to exhibit considerable anti-inflammation osteo-arthritic effects, enhance the growth of bone collagen and to effectively protect the degradation of cartilage in studies conducted on mice (190) (191) and Ginsenoside Rg1 may have the ability to protect the lens of the eye against cataracts caused by oxidative stress (192) (193). Ginsenoside Rb1 also protects against Hydrogen Peroxide-induced Oxidative Stress in studies conducted on mice (194).

Sun Ginseng

Sun Ginseng is a heat-processed type of ginseng which has been shown to exhibit anti-obesity effects (195). Sun Ginseng is made by steaming raw ginseng at a specific temperature (195). During this special heat processing method, Sun Ginseng generates above average levels of the ginsenosides Rk1, Rg3 and Rg5, compared to red and white ginseng varieties (195).

Getting back to the original study stated earlier (174), researchers than decided to examine the effect of mixing the amino acids leucine and arginine with ginseng to examine how well it fought cancer. What is interesting is that after the heat processing with the amino acids leucine and arginine, which created the Maillard reaction in the ginseng, the new ginseng substance was found to contain glucopyranosyl and rhamnopyranosyl. As will be shown later, leucine, when it is heated to extreme temperatures, actually causes problems, such as the leucine in meat. It is key that the leucine is combined with the right substances and at the right temperature in order for it to exert its anti-aging and therapeutic properties (196).

Rhamnose

Rhamnose has been shown to accelerate the healing of skin wounds up to 10 days faster, compared to the control group (197). Rhamnose is also created in the body by the stomach's microflora after consuming ginsenosides (198) (199). This is because ginsenosides have been found to contain rhamnopyranosyl (200). Rhamnopyranosyl is found in Moringa oleifera leaves (201) and Rhamnopyranosyl. Acacetin 7-O-α-l-rhamnopyranosyl (1–2) β-D-xylopyranoside has been found to extend the lifespan of worms more than 39% (202). Acacetin 7-O-α-l-rhamnopyranosyl (1-2) β-D-xylopyranoside is also found in the edible leaves of Premna serratifolia (203). What is most fascinating is when Premna serratifolia leaves are made into an extract, it protects DNA. The researchers of the study also stated that the DNA protection observed may be one of the major factors that contribute to its longevity and its anti-aging effects [203]. The extract also contains Acacetin 7-O, which I shall show later on is a powerful anti aging substance.

Rhamnose is found in Sea Buckthorn, poison sumac and Uncaria genus plants and can also be produced by microalgae in the Bacillariophyceae class. Rhamnoides from Sea Buckthorn has been shown to significantly increase the lifespan of worms up to 29.40%, while acting as a neuroprotective agent [204] and when olive oil was mixed with Sea Buckthorn, it was found to be useful for treating burns [205]. Rhamnopyranosyl has also been found to protect DNA in research studies conducted as far back as 2016 [206]. Wax gourd (Benincasa Hispida) has also been found to contain not only rhamnopyranosyl, but the potent anti-aging substances oleanolic acid (*found in olive oil*), stigmasterol and nicotinic acid [207], with the dried fruit pulp being recommend as an anti-aging preparation for reducing the symptoms of skin aging [208].

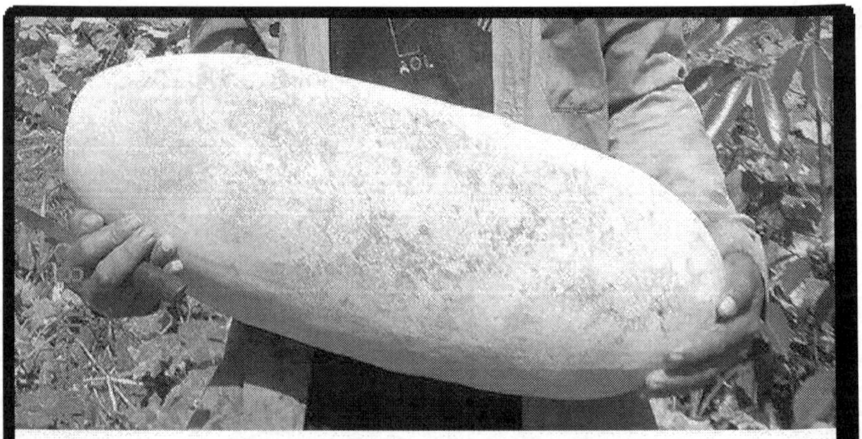

Benincasa Hispida

Chapter 7
Exploring the Anti-Aging Potential of Rhamnopyranosyl

Doug Skrecky (www.cryonet.org) is famous for his hundreds of longevity experiments on fruit friles. Out of his hundreds of longevity research experiments, he discovered the food Wax Gourd (Benincasa hispida) and standard coffee extended lifespan the most (209) (210). Wax Gourd is also called Kundur. An extract made from Wax Gourd peels has been researched for its ability to prevent obesity and treat hyperlipidemia (211). Wax Gourd Juice has also been shown to help create good blood circulation in the veins and arteries (212) and its seeds are rich in antioxidants (213). When we look at an anti-aging study (214) regarding Wax Gourd (*Benincasa Hispida*), when the fruit pulp was dried and tested as an antioxidant and skin renewal enhancer, it acted as an effective skin anti-aging preparation, with the researchers recommending it as a substance that can be used to retard the symptoms of aging. In a recently study published study (June 2022) (215), researchers found Wax Gourd exhibited favorable anti-aging properties and suggested it for use as an effective substance for developing anti-aging skin care products. Wax Gourd has also been used to treat fever, coughs that are accompanied by thick mucus, intestinal worms and neurological diseases and kidney diseases (216) (217) and an extract of Wax Gourd has been shown to exert antioxidant activity (218). Wax Gourd has also been shown to be a good source of amino acids, organic acids, vitaminis, natural sugars and minerals (219).

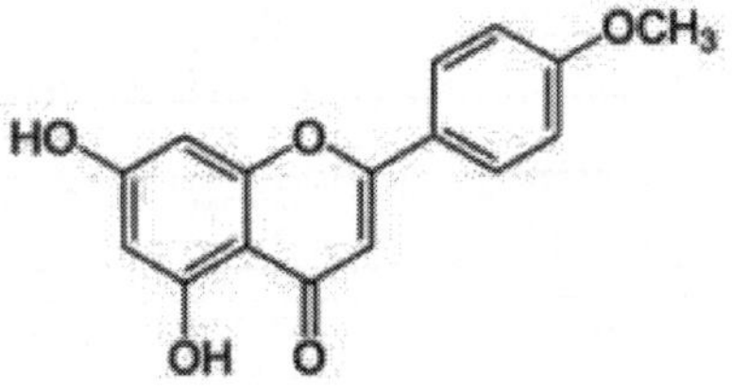

Acacetin

When Acacetin 7-O-α-L-rhamnopyranosyl was isolated from Premna integrifolia, it was found to extend the lifespan of worms up to 39% (220). The study found the effects of the rhamnopyranosyl based substance mimicked dietary restriction. The authors in the study concluded that this phytomolecule is recommended as a therapeutic for managing age-related diseases. Premna integrifolia powder or capsules are sometimes called Agnimantha, and are commonly sold as a supplement online. The parts of the plant most commonly used are the roots, bark and leaves.

Rhamnopyranosyl as an airborne pollutant detoxifier
Rhamnopyranosyl may also help fight the effects of air pollution. A study (221) found that an extract of Camellia Japonica flower was able to act as an active cosmetic component to help fight the effects of skin aging in environments that have air pollutants that contain heavy metals. The study also found that it induced a slight increase in collagen density and when the extract was tested in an environment without pollution, it was found to significantly reduce oxidative stress by 16%. This is a significant discovery, because some of the best anti-aging substances that detox pollution from the body are also some of the most effective anti-aging substances. See Chapter 17 for a list of the best substances for reducing the body's exposure to air pollution.

Camellianoside
The leaves of Camellia Japonica flower contain an abundance of the flavonol glycoside known as Camellianoside. The structure of Camellianoside includes quercetin, rhamnopyranosyl and xylopyranosyl (221).

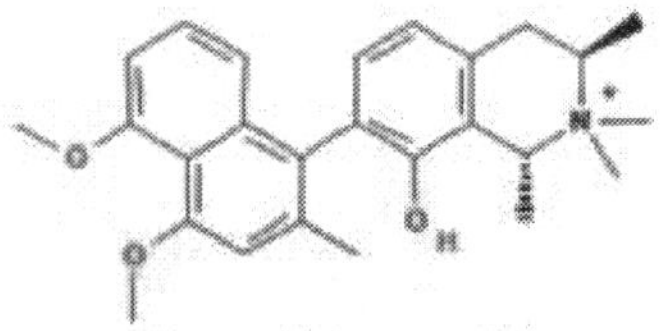
Camellianoside

Camellianoside is a powerful antioxidant, having an antioxidant superiority similar to l-cysteine and vitamin C (Onodera et al., 2006). Extracts of Camellia Japonica flower were found to have the following properties – the chelation of metals from the body, the elimination of free radicals, ROS quenching, bacteriostatic activity against a wide range of food borne bacteria, antifungal activity, reduction of oxidative stress and a decrease in glucose blood levels (223). Camellias are also used to produce the edible oil oleic acid (Su, Shih, & Lin, 2014).

Camellia Seed Oil contains palmitic, oleic acid and linoleic acids and oil from its seeds contain up to 80% fatty acids (224). This is a very high amount of oleic acid, similar to the oleic acid levels found in Olive Oil (225) (*Olive oil contains between 5 and 80% Oleic Acid*). Studies have confirmed that the oils of Camellia Japonica contain a structural make-up similar to that of olive oil (Fang, Fei, Sun, & Jin, 2016). However, the oils in Camellia Japonica are lower in saturated fatty acids (Salinero & Corral, 2008), which makes it more healthy (226).

Chapter 8
Exploring the life giving power of Gotu Kola (C. asiatica)

The herb known as Centella asiatica is also known as Indian Pennywort, Jal brahmi or Gotu kola. It is part of the Umbelliferae species and is used in Ayurveda to treat age related brain disorders and enhance memory [227]. It has also been found to enhance the health and growth of brain neurons and to help with nerve regeneration [227]. It also contains powerful antioxidants and essential oils from its leaves contain pinene, one of the most powerful anti-aging substances known.

When steam or water was used to extract nutrients from the leaves of Gotu Kola, **it was found to contain Carene** [228]. As covered in an earlier chapter, when ginseng is heated with amino acids, it produces Rg3 and Rg5. It just so happens that Rg3 and Rg5 are also found in Gotu Kola [229] [230]. As stated earlier, besides consuming rice wine and goji berries, Gotu Kola is reputed to have been consumed by Mr. Li Ching-Yuen who allegedly lived to 256 years of age [231]. For now, the science regarding the anti-aging substances in his diet do seem to substitute the claims that he did indeed live for over 200 years.

When Gotu Kola is fried, its polyphenol content is reduced to only 20.6% [233]. This is an interesting finding because out of all plants that underwent frying in this study, Gotu Kola showed the strongest resistance to the loss of polyphenols during the extreme heat. In other words it showed the lowest loss of polyphenols during the frying. This proves Gotu Kola has powerful heat shock protection properties.

SOX9 activation and Gotu Kola

Gotu Kola is also abundant in asiatic acid [234]. Asiatic acid has been shown to boost activity in the SOX9 gene **up to 151.04%** (235). Asiatic acid also protects cartilage by promoting chondrogenesis (*the development of skeletal bones*) and also inhibits osteo-arthritic inflammation (235). (Zhengmeng Yang et al. Sept 2021). Gotu Kola also contains the terpenoids madecassoside and asiaticoside, which is proven to increase collagen levels (236) and exhibit anti-aging effects. So now we have powerful factual data that supports Gotu Kola is a powerful anti-aging herb. Now how do anti-aging herbs affect the genes that extend lifespan? Let's dive into the science some more to seek the answers.

Upregulation of the SOX9 Gene

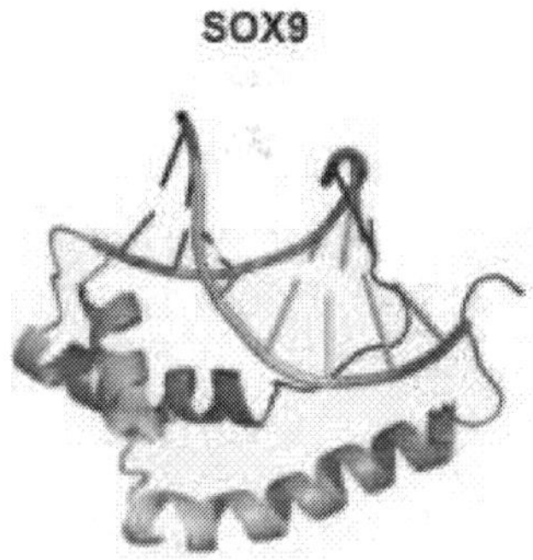

What is all this hype about SOX9? To put it in simple terms, SOX9 is a gene that helps create collagen and maintain cartilage integrity (237). One of the major hallmarks of having healthy levels of SOX9 gene activity is that it will reverse the effects of osteoporosis and restore bone mineral density (238). In studies on mice that were specifically bred to express an over-abundance of the SOX9 gene were found to have avoided senescence (239) and in SOX9 stem cell research studies (240), SOX9 was found to promote epithelial regeneration (cells which line the external and internal surfaces of the body).

When the SOX9 gene is stimulated to above average levels, this upregulation will resonate for up to 7 days thereafter (241). There is also good scientific proof that SOX9 helps maintain healthy bones. For example, when SOX9 levels are low, it has been shown to affect articular cartilage (242), which is the white elastic tissue which allows your bones to glide smoothly together (243).

Because bone degeneration is one of the major factors of aging, for the most part, any food or substance which upregulates / stimulates SOX9, extends lifespan and in some cases reverse aging. Science has shown that the most powerful SOX9 upgregulator is Baicalin, which is also one of the most powerful anti-aging supplements (244). Baicalin is found in the plant Scutellaria baicalensis (*also called Skullcap*). Chinese Skullcap roots contain diterpenoids, which have significant antioxidant activity (245). Diterpenoids are used for treating gout, rheumatism, palsy and amenorrhoea (246). Diterpenoids can be found in marine sponges, fungi and bacteria (247). Diterpenoids also include ajugarins, which are repellents that repel herbivore predators, with some tasting very bitter, such as gymnocolin (247).

SOX9 Gene Stimulators -
Angelica sinensis polysaccharide, malvidin, chrysanthemum, psoralen (found in fructus psoraleae), danggwisu-san, scutellarin, silymarin, broccoli extract, daidzein, oleic acid (*found in flaxseed oil, sunflower oil & olive oil*) and icariin, which is isolated from horny goat weed and the epimedium plant epimedium pubescens.

Other Sox 9 Stimulators include (248):
Kaempferol, apigenin, rhizome of ligusticum chuanxiong, and schisandra chinensis, which also helps protect the skin against air pollution (*Schisandra chinensis Protects the Skin from Global Pollution. Edwige Ranouille et al. June 2018*).

Atractylenolide (249) (found in the rhizomes of A. macrocephala), Isopsoralen (250), UP256 containing bakuchiol (251).

Nutrient Depreviation
When mesenchymal stem cells are deprived of nutrients for 24 hours, they cause cells to rapidly adapt to nutritional starvation, which in turns increases SOX9 levels (252).

What is SIRT1 and how does it work?
SIRT1 is a gene that protects your brain from neurodegenerative diseases, neuronal damage and your heart from atherosclerosis (253) (254) (255). When people are first introduced to the anti-aging

literature, they believe that enhancing SIRT1 extends lifespan. However the evidence is conflicting. For example, in one study (255) mice having an overexpression of SIRT1 that were fed a standard diet, did not live longer lives. However what was interesting was that the mice enjoyed better overall health and were less susceptible to cancer. This research was on mice fed a standard diet. Vegan or Vegetarian diets with enhanced SIRT1 levels may allow people or mice to live longer lives, although further studies are needed to confirm this hypothesis. A research study published in February 2022 titled: Risk of cancer in regular and low meat-eaters, fish-eaters, and vegetarians, that was authored by Cody Z. Watling and colleagues, concluded that vegetarians were less likely to contract many forms of cancer including breast and prostate cancers.

The SIRT gene has been found to interact with DNA (Sinclair and Guarente, 1997) (256) and studies hypothesize that activation of this gene may play a role in extending the lifespan of yeast, but not its chronological lifespan (Kaeberlein et al., 1999; Fabrizio et al., 2005). Regarding yeast, lifespan extension via upregulation of the SIRT gene was found to extend lifespan by mimicking the effects of caloric restriction (256) (Lin et al., 2000; Anderson et al., 2003; Cohen et al., 2004). In other studies an over activation of SIR2 was shown to increase the lifespan of flies and worms (257) (Tissenbaum and Guarente, 2001; Rogina and Helfand, 2004; Wood et al., 2004).

In a recently published study (258), SIRT1 levels were higher when non-vegetarians exercised and SIRT1 activity decreased in vegetarians. What was also interesting was that when the lacto-ovo vegetarians exercised (*a diet that is mostly plant based, but without poultry and meat but includes eggs and dairy*) their SIRT1 levels increased. Regarding the vegetarians, this could be because plant-based diets are usually low in vitamin B12, which may reduce SIRT1 levels (259). It is interesting to note that fhe most common vitamin deficiency in the global human population is Vitamin B12 followed by Vitamin A (Revisiting food-sourced vitamins. Chigozie E. Ofoedu et al. Sept 2021). This is expected to grow as more and more people switch to plant based diets.

Low levels of SIRT1 have been shown to reduce SOX9 functioning, with a research study confirming that SIRT1 activates SOX9 (260). SIRT1 can also be increased by prolonged fasting between meals or intermittent food availability (261). Intermittent fasting studies (262) involving fruit flies, yeast, non-human primate animals and mice show that Intermittent Fasting not only activates SIRT1, but also increases NAD+ levels, delays senescence, extends longevity and delays the onset of disease.

Chapter 9
Exploring Alchemy, the Red Powder and the Philosopher's Stone

The red powder is a substance that allowed one to transmute metals. It was sometimes called the "*Powder of Projection,*" allowing any metal to become transmuted into silver or gold when wrapped in a ball of wax and added to molten lead (263). The dry red powder was also called al-kibrit al-ahmar, red sulfur and was perfected by Arab alchemists (263). What is most interesting is the two most powerful anti-aging substances quercetin and baicalin, interact with lead, by attaching itself to the lead so as to act as a metal chelator or transforming it (263) (264). It has been hypothesized that the first Alchemists originated in Ancient Egypt. Chinese writings on Alchemy were first recorded around 100-150 B.C.E. This was long before Alchemy was being experimented with in Western Europe. The Chinese were the first to use Gold for treating illness and for longevity. It is reported that Wei Po-Yang was able to make the red powder and that it was the source of his immortality (265).

Sabine Baring-Gould stated that the effects of the Philosopher's stone could be traced to the life-giving properties of the sun (265). The sun, including that of the constellation Leo, is a popular symbol in much of the alchemical literature. He wrote about this concept in detail in his book titled: Curiosities of Olden Times (1895).

It is my hypothesis that there exist numerous substances that can be used to make the red powder. However the one that has gained the most attention is Urine. Inside the London British Museum, a middle of the XV century manuscript titled: Cabala Mineralis or the She Horse on Urine Work, written by Rabbi Simeon Ben Cantara goes into great detail about this. What is most interesting is urine from cows is used to make what's known as a "bioenhancer." A bioenhancer is any substance which accelerates plant growth and increases crop yield. This means that certain bio-enhancers may possibly be used to create the red powder. Indeed a research study found that cow urine exhibited significant effects upon the production and growth of sorghum plants (266) and in another study titled: Cow urine distillate as bioenhancer, the researchers stated cow urine (*also called gomutra*) has been documented in Ayurveda texts as "Ashtanga Sangraha" and "Sushruta Samhita" as having innumerable therapeutic properties. Gomutra is also used to purify bhalataka (Semecarpus anacardium), guggul (Comniphera mukul), loha (iron), detox aconite (Aconitum napellus) and to detox and purify silver.

Cow urine also contains the following amino acids, which are listed in order from the highest to the lowest: (267) **Glycine**, Serine, Alanine, Leucine, and Proline. This is an interesting finding because grass dew also contains the amino acid proline (278).

In the text titled: Bhav Prakash Nighantu, it describes gomutra as being the best out of all types of urine from all the animals (*including human urine*) (269). People who drink gomutra were said to live healthy, remaining healthy well into their 90's (269). Gomutra has found applications as a bio-pesticide in organic farming and some poisons are purified and refined when soaked in gomutra for a total of three days (269). The poisonous Dhatura plant, including its seeds (Datura stramonium) with the shells peeled off, are said to become purified after soaking them in gomutra for 12 hours (269). This is a major discovery as the only substance that also purifies toxins when substances are soaked in it is coriander. Extracts of coriander leaves have been used to reduce the contamination of heavy metals such as lead, mercury and copper that exist in shellfish and fish (270). Gomutra has also been shown to act as a strong bioenhancer (*or to amplify the effects*) of the prescription drug Rifampicin, which is used to treat tuberculosis. The research showed a sevenfold increase in its fighting activity against Escherichia coli as well as an 11-fold increase in fighting

Gram-positive bacteria, one of the main class of bacteria responsible for tooth decay.

Gomutra has also been shown to enhance the transport of the prescription antibiotics Ampicillin, Rifampicin and Tetracycline (269) and it also exhibits antitoxic activity against cadmium chloride (269) and is used as a bioenhancer for zinc (269). When male mice were exposed to cadmium chloride, they exhibited a 100% infertility rate. However, when the mice were given a combination of cow urine and cadmium chloride along with zinc sulfate, they demonstrated healthy rates of fertility (271). Cow urine has also been granted US Patents (No. 6,410,059 and 6,896,907) for use as a bioenhancer, antibiotic, anticancer and antifungal agent (272). It has also been issued a patent for use in increasing the potency of Taxol, which is used to treat human breast cancers (US Patent No. 6,410,059) (272).

Creating Gold from Urine

One recent study only published a year ago titled: A comparative study on anti-microbial efficacies of biologically synthesized nano gold using Bos taurus indicus urine, that is published in Current Research in Green and Sustainable Chemistry (Volume 5, 2022) and was authored by C.A. Pawar and colleagues, stated that they were able to make gold from urine of the Indian Gir Cow. The process involves a heat method and the resulting gold also shows powerful antibacterial properties against the following bad bacteria (shown from highest to lowest) - Staphylococcus aureus, Aspergillus niger and Pseudomonas aeruginasa.

Urine enhances Crop Yields

Studies using human urine as a fertilizer, found that the yields of Swiss chard were four times greater compared to the control group (273). Out of all natural fertilizers tested, human urine gave one of the highest yields (274), which was followed by dried chicken manure and a combination of meat and bone meal. Could the reason the urine enhances crop yields be due to the fact that heat from the sun and other environmental forces interact with the urine to create beneficial substances?

When you think about it, when the body makes urine, part of the process involves a long warm steady extheromic type process. This process is discussed further in this book when bacon is mixed with wood chips to create numerous anti-aging substances. As wood chips decay, they undergo the exothermic nature of pyrolysis (self heating).

Another bio-stimulant which uses a mild heated chemical reaction in water is when Lye is mixed with Celtic sea salt and the pH is kept at a slightly alkaline pH level. The resulting substance is Ormus (Orbitally Rearranged Monoatomic Elements). See my book titled: The Official Guidebook of How to Make Tinctures and Alchemy Spagyric Formulas, for detailed instructions on how to make your own Ormus. Duckweed, which is a plant byproduct that thrives in wastewater is used as a bio-stimulant (275) and Duckweed contains the amino acids valine, leucine, threonine and phenylalanine. It also contains above average amounts of lysine (276).

A research study (277) found that the diets of milk-fed young pigs are very low in arginine. However when the ratios of lysine to arginine was increased in their diets, it resulted in reductions of ammonia in their blood and they experienced more healthy weight. Hence lysine is acting as a detoxer of waste and toxins. For example, a research study found that magnetic nanostructures with lysine removed heavy metal pollutants from sludge (278).

Since the commercial advent of Ormus almost 15 years ago, it has since yielded thousands of testimonials of its ability to create large crop yields and improve overall health and well-being when taken orally. In summary, because urine is one method to make the Philosopher's Stone, if one could understand the biochemical process of how urine is formed in nature, one could create a powerful healing and possible anti-aging substance and avoid the negative stigma associated with using urine for therapeutic purposes.

Using the Red Powder

It is my hypothesis that the red powder of the Philosopher's Stone acted as a type of prebiotic that enhanced the bioavailability of the probiotics in the red wine. This means that adding the sacred red powder to any type of probiotic such as yogurt or kefir would allow it to unlock its life extension capabilities. It is also interesting to note here that when a small amount of the red powder, equal to a grain of rice, was wrapped in a ball of wax and added to liquid lead, allowing the red powder to gradually heat up, it would turn lead to gold. Lead is one of the most dense metals in the element kingdom. Hence the red powder merely enhanced its bioavailability, allowing it to penetrate the lead in order to transmute the lead into gold.

Only a few grains of red powder were added to a bottle of red wine, and then only a glass or two was drank each day to give the drinker physical immortality. It is my hypothesis that when the red powder was added to red wine, that it stimulated a mild Maillard reaction in the wine (perhaps the probiotics in the wine) which enhanced the antioxidants, probiotics and other beneficial substances in the wine, due to the red powder making the red wine more bioavailable by the body.

I have created a unique prebiotic (*a room temperature substance that enhances good bacteria in the stomach*), that enhances the probiotics when added to yogurt or kefir. It consists of 1 part coriander, 1 part astragalus and .05 parts olive leaf powder. This is then mixed with 1/2 teaspoon of cinnamon and added to yogurt. If you make and take this you can really feel the difference after adding this to yogurt, especially if you allow the powder to sit in the yogurt for 10 to 20 minutes before eating. Indeed studies show [279] that when prebiotics are combined with probiotics, that it may increase the efficacy of the probiotic. And peanut sprout yogurt exhibits more antioxidants compared to normal plain yogurt [280]. This is because peanut sprouts contain an abundance of polyphenols and resveratrol (Yu et al. 2022).

It is now a confirmed scientific fact, confirmed by numerous published studies over the last few decades, that red wine contains numerous anti-aging substances. For example a research team (281) had practicing religious Nuns drink red wine in moderate amounts to see what effect it had on their longevity genes. The genes were SIRT1, p53 as well as SOD levels (which fights free radicals). The research team found that these genes showed increased activity after the Nuns' moderate consumption of red wine for 14 days. The red wine consumption also upregulated expressions of the gene 16S rRNA, which decreases in activity as a person ages. There were also significantly lower levels of the amino acids leucine and isoleucine as well as an increase in Taurine levels in the body. As covered in Chapter 3, the amino acids leucine and isoleucine are problem amino acids that possibly shorten lifespan (92). The study also cites studies which show that drinking red wine significantly increased good cholesterol levels in the human body, but not by drinking white wine. Studies also found that red wine also increases the lifespan of fruit flies (281). Other studies confirm (282) that the moderate drinking of red wine attenuates dementia, helps diminish the deterioration of Alzheimer's disease, increases cellular viability and helps improve memory. In Ayurveda, fermented red grape juice is used to treat heart conditions (282) and in the Bible, Timothy drank sparse amounts of wine to alleviate his stomach and frequent illnesses. In 1992 Siemann and Creasy confirmed resveratrol was the active ingredient in wines (283) with the resveratrol having a half-life of between 2 and 14 hours in the body after taking it (283). White wine contains significantly less resveratrol compared to red wine (Resveratrol in Wines and Grapes. Dr. Liji Thomas). Indeed from my own personal experience of opening up a resveratrol supplement capsule and pouring 1/3rd of the contents of the capsule into hot boiling water, than adding stevia, allulose or honey to sweeten than drinking it as a tea, significantly enhances my memory after taking it for 3 days in a row.

Exploring the Mysteries of the Polyphenol Resveratrol

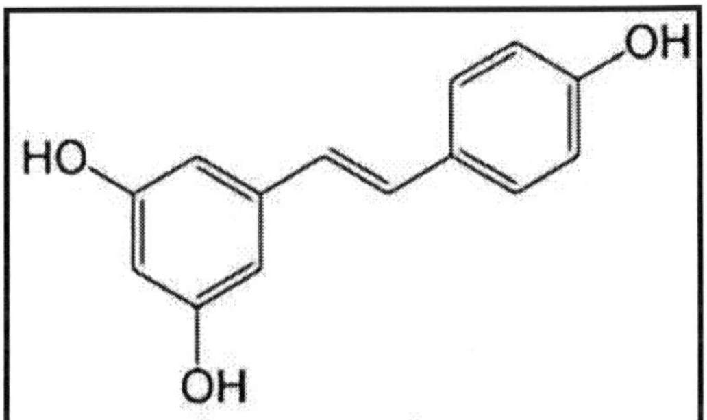

Resveratrol

Resveratrol was first isolated in 1940 from the roots of white hellebore (Veratrum grandiflorum O. Loes) and then from the roots of Polygonum cuspidatum (also called Japanese Knotweed) in 1963. In nature, Resveratrol is produced in response to stress conditions experienced by plants (285) (*and possibly mammals*) and is found in large amounts in grape skin, raspberries, peanuts, pistachios and mulberries as well as 72 edible and medicinal plant species (285). In grape skins, resveratrol represents 5 to 10% of the skins and exists at between 0.05 and 25 mg/L in grape juice and wine (286). In pistachios, peanuts, blueberries and bilberries, its concentrations can reach up to 5.1 µg/g with the plant species R. Rhamponticum containing approximately 3.9 mg/g (286).

In closing, it is interesting to note that Hindu mythology states that cows are sacred animals and the religious Black Stone, which is set into the Kaaba in the the Grand Mosque in Mecca, Saudi Arabia, is said to be a meteorite. This stone is worshiped by Muslims. Both of these substances, urine from cows and meteorites have relationships to gold. An August 2016 article published by chromatographytoday.com titled: How Are These Cows Literally Peeing Gold?, found that one litre of Gir cow urine can contain up to 10 mg of gold.

Chapter 10
The Life Giving Powers of Arginine and Glycine

Just as glycine protects the body from toxic levels of methionine (112), a research study (287) involving rats found that when arginine was added to a mixture of 9 amino acids, it caused a reduction in mortality from 100% to 24%. The favorable effect arginine has on the body may be from physical exercise. This is because as shown earlier in Chapter 7 that the glycine in the Ginseng, when heated, results in an increase in the anti-aging substances in the Ginseng.

When athletes took a combination of arginine and citrulline for a number of days, they showed improved physical performance and experienced less fatigue (288) and mice studies have found that arginine preserves exercise performance capacity and reduces muscle fiber damage (289). Another study (290) found that arginine enhanced mitochondrial functioning and reduced oxidative damage in the brain. The authors of the study concluded that taking arginine daily slowed aging and reduced stress-induced brain damage. When glycine is heated to 240 degrees Celsius, it can form aspartic and glutamic acids (291) and when yeast is fed a methionine restricted diet, with methionine levels being reduced 10%, and the diet is also low in glucose and high in glutamic acid, the yeast will live 2.3 times longer (293). What is interesting is that an extract of Centella asiatica (gotu kola) has been found to stimulate glutamic acid decarboxylase functioning 40% or more (293). This is an enzyme responsible for the catalyzation of glutamate to GABA, which is the main neurotransmitter of the human central nervous system. Both aspartic acid and glutamic acid has been used to protect baby pigs against nutritional deficiency (294). It is interesting to note here that when pigs were put on a fast, that they exhibited higher levels of alanine, aspartic and glutamic acid (295) and a combination of aspartic acid and glutamic acid have been found to be a good combination to help fight cancer (296).

Chapter 11
Digging into the Longevity Secrets of the Bristlecone pine tree

The Bristlecone pine has been scientifically verified to live up to 5,000 years (297). Just as the famed Chinese immortals were reported to live high up in the mountains, the Bristlecone pine tree lives high up in the mountains in an environment with scathing winds, acidic soil and long cold harsh dry winters (298). This cool, dry environment with low rainfall means it exists in an environment that has high hydrophobicity, where dust and dirt cannot cling and adhere to the tree for long periods of time and as shown in this book, an overexpression of the SOX9 gene has been used to heal numerous artiritus, osteo-arthritic and cartilage disorders. It is interesting to note that the gene level of SOX9 is significantly down regulated in people diagnosed with osteoarthritis compared to people who have not been diagnosed with osteoarthritis (299). We shall cover why there is a connection between this and the Bristlecone pine tree a lttle later on.

Bristlecone pine tree

Because Pinene has been studied for its ability to prevent osteoporosis, it makes sense pinene may be a possible candidate for SOX9 gene upregulation (300) and also as shown in this book, many anti-aging substances produced in plants are used to repel pests and herbivores (an animal that only eats plants). For example, terpinolene and limonene possess potent insect repellant properties (301). I hypothesize that the reason this pine tree lives so long is because the harsh environment forces it to produce substances to protect itself, which in turn creates substances that reduce the aging process. Evidence to support this hypothesis is the fact that one of the most powerful anti-aging substances, resveratrol, is made in over plant 70 species as a response to stress in its environment (302). This is similar to the process where a sheep or horse is injected with just enough snake poison to not kill the sheep, but to allow the sheep to produce antibodies which are used to treat snake bite (171).

The role Phytoalexins play in nature

Phytoalexins exist in nature as antimicrobial compounds with a very low molecular weight. Resveratrol just happens to be a phytoalexin (*stilbene*), which also possesses very promising antioxidant properties (303). The phytoalexin Resveratrol is made by over 70 plant species in response to stressful situations (304). Phytoalexins build up slowly over a period in time in plants and accumulate as a byproduct of their resistance to adverse weather conditions, parasites, fungal attacks, UV radiation, chemicals and general plant stress.

The Bristlecone pine tree must be making a type of Phytoalexin. Research shows that the most abundant substances in the Bristlecone pine are Carene, Pinene, Limonene, Phellandrene, Terpinolene, Camphene and Myrcene (305) (306). Carene just happens to be a substance emitted by plants in response to insect attack (21).

Carene has shown extremely potent antibacterial activity against the gram positive bacterias P. fluorescens and B. thermosphacta (*bad bacteria*) (307). The effect was so strong it even caused the death of the bad bacteria right down to its cellular level. Besides the Bristlecone pine tree, Carene is also found in white and black pepper oil (307).

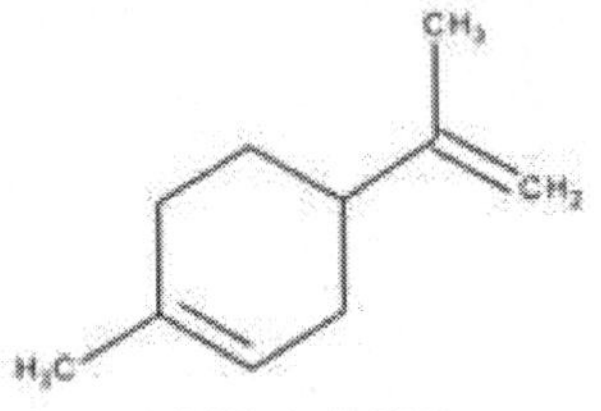

Limonene

Carene is often found in nature with pinene and limonene. What is most interesting is when limonene is added to the Bristlecone pine tree, it disrupts a beetles ability to distinguish between species of trees [308]. This effect was not present when the limonene was added alone.

Uncovering the hidden secrets of Pinene

The Bristlecone pine tree has been found to have significant amounts of the monoterpenes Pinene, Limonene and Myrcene [309]. These monoterpenes have been used to reduce the effects of Heat Shock Protein. For example limonene has been shown to inhibit heat shock [310]. I shall cover later in this book how the prevention or reduction of heat shock proteins extends lifespan and creates a healthier life. Pinene protects the skin against UVA rays as well as exerts strong skin protection against ROS [311]. Pinene has also shown greater antioxidant activity than vitamin C (ascorbic acid) [311].

Pinene's power comes from its ability to easily cross the blood brain barrier where it exerts its neuroprotective and anti-depressant effects through its terpene like odor [312] and when pinene is combined with linalool, the combination exerts a synergistic effect [313]. β-pinene, targets the brain's serotonin receptors and exerts antidepressant effects [314].

What is the Difference between α-pinene and β-pinene?

A-pinene, has a pine-rosemary type scent and dissolves more easily in water compared to β-pinene, which has a parsley basil type scent and dissolves more easily in fats. Studies on mice also found that pinene may improve sleep simply by inhaling pinene before bed and other mice studies showed Lemon essential oil (which contains pinene) chelated metals [315], which is the ability of a substance to remove toxic metals out of the body. Because pinene has such a low molecular weight and high lipophilicity (easily enters fats and oils), it rapidly penetrates the blood brain barrier. Mice studies [316] have shown that after breathing in pinene, that the pinene was detected in the brains of mice for up to 30 minutes after and the same research study concluded that levels of α-pinene were 2-fold higher when the pinene was delivered in an oil blend. This was done by allowing pinene to circulate in the open air when it was mixed with limonene, cymene or cineole. Pinene is also used as an antidote for aspirin overdose (317) as well as inhibit ROS production, and significantly enhance antioxidant enzymes responsible for fighting the aging process (318). These antioxidant enzymes include glutathione peroxidase (GPx), glutathione reductase (GR), superoxide dismutase (SOD) and catalase (CAT). The glutathione gene is a gene that helps the body remove environmental pollutants (319). Superoxide dismutase (SOD) is often called the most important antioxidant made in the human body because it acts as a front line defense against inflammation and damage from free radicals.

Pinene has also been shown to exert neuroprotective effects (320). and has been found to act as a powerful natural antioxidant compound (321). Besides its neuroprotective benefits, pinene acts as an anti-inflammatory, protecting the brain against neuronal damage (322) and pinene enhances memory recall in mice studies, where mice showed improved spatial recognition and increased learning (323). The study also found that pinene also enhanced antioxidants in the brain's hippocampus.

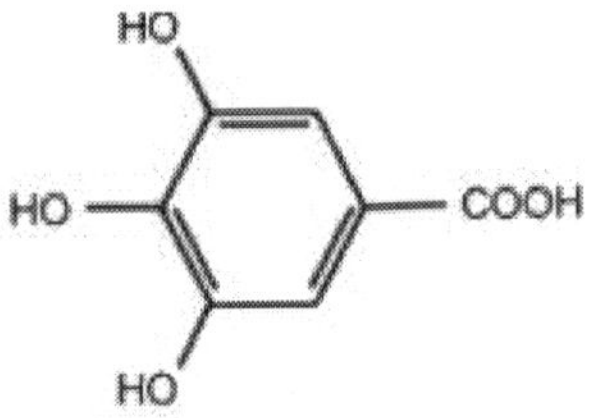

Gallic acid

Pinene is formed when Peony seeds are heated with chicken fat and the amino acid cysteine (324) and Moutan Peony flowers (*Paeonia × suffruticosa*) contain geranial, gallic acid, apigenin-7-O-glucoside, linalool and carene (325) (326), all powerful anti-aging substances.

Natural oil extracted from Peony seeds has been shown to significantly reduce many of the components responsible for aging (*malondialdehyde and monoamine oxidase*) and has been shown to increase levels of Glutathione, Manganese Superoxide Dismutase (MnSOD) and Cu,Zn superoxide dismutase (SOD1, all of which exhibit anti-aging properties, as well as eliminate the aging effects of the brain sugar d-galactose in mice [327].

Carene can be obtained from -
Culantro (Eryngium foetidum), which contains 16.19% carene and 25.42% limonene.

Mango peels contain carene [329] [330] and pinene. And mango peels contain more carene in the skins, compared to the pulp; accounting for over 80% of the aroma in the peels. Treatment by ethylene caused carene formation in the mango peels at the start of storage to reach maximum levels on the third day [329] [330]. Treatment of the peels using the bioenhancer methylcyclopropene reduced carene levels during storage, however it increased carene levels in the mango pulp [329].

Pinene and its effects on the body
Plants abundant in pinene have been used in traditional medicine for centuries to treat rheumatism, cancer, anxiety, depression, gastrointestinal disturbances, seizures, inflammation, pain, colds and snake bite, fevers, hypertension and fungal infections [331].

When the essential oil α-pinene was obtained from Ugni myricoides leaves and orally given to mice at 5 to 25 mg/kg, it exhibited anti-inflammatory and analgesic effects against carrageenan toxins [332]. Essential oils made from Shell ginger (Alpinia zerumbet) and Moshgak (Ducrosia anethifolia) have been shown to reduce anxiety and increase the levels of α-pinene in the brain after inhaling the essential oils [333] and inhaling the essential oil of Hinoke Cypress (Chamaecyparis obtusa) for an hour and a half showed that the molecules of the oil gathered in the hypothalamus and pre-frontal cortex regions of the brain where it than exhibited a reduction in anxiety [334]. The hypothalamus is a part of the brain that is the arbiter which orchestrates systemic aging through neuroinflammatory signaling [335].

In a similar research study with mice, breathing in Roman chamomile (Chamaemelum nobile), which is abundant in α-pinene, reduced depression (336). Further examination revealed that the a-pinene increased mitochondrial functioning in the brain's pre-frontal cortex as well as hippocampus. The hippocampus is the part of the brain that is a critical structure related to cognitive aging. It affects processing speed, memory and executive brain function (337). Pinene also reduces migraines, neuropathic pain and improves cognition (338). Several research studies confirm that pinene exerts similar or even better healing benefits than some prescription medications (339). The common household Lemon contains an abundance of natural pinene (α-pinene and β-pinene) as well as linalool, limonene and caryophyllene (341). Lemon also contains carene (342). These substances are all powerful AGE fighters. This is why adding lemon to high AGE foods significantly reduces the negative effects of AGE's (342). Besides Lemon, out of many species tested for carene levels, the leaves of Bitter Orange were found to contain limonene, pinene and carene as the major compounds (343).

Mexican Bay Leaf (Litsea glauescens) essential oil contains an abundance of linalool and β-pinene and has been used to alleviate depression and anxiety, as well exhibit a 2 fold pronounced effect on uplifting moods; more than the prescription pharmaceutical medicine imipramine (344). An abundance of α-pinene is found in CBD Skunk Haze and Lemon Skunk (345). β-pinene is also abundant in Purple Kush (345). Pinene is also abundant in turpentine, lavender (Lavandula stoechas) and rosemary (Rosmarinus officinalis) (345) and a-pinene occurs in frankincense (Boswellia frereana) (346).

A clinical trial (347) involving twenty elderly men with an average age of 60 years were given frankincense supplement pills to be taken twice daily (500-mg). The study found that the frankincense supplement exhibited significant positive effects upon their memory and their ability to retain information.

In another study (348) that looked at the main compounds in Frankincense from Ethiopia, the researchers found that out of 29 substances from the leaves, that α-pinene (45.7%) and carene

(27.7%) were the dominant substances.

Frankincense also affects the brain's hypothalamic pituitary region, helping the body better adapt to stress [349].

β-pinene is found in abundance in Ferula gummosa, rubricaulis, balsam fir (Abies balsamea) and Guinea grains (Aframomum angustifolium) [350]. Greek sage (Salvia Fruticosa) which contains an abundance of pinene, has been shown to prevent oxidative stress [351]. The essential oil from Spanish sage (Salvia lavandulaefolia) exhibits signifiant anti-inflammatory activity and antioxidant effects in mice leukocytes (*a cell in human blood which helps the body fight diseases*) [351].

Chapter 12
Secrets of the life Giving Powers of the Anthocyanin

Could there exist a synergy between Anthocyanins and Carene? Dendrobium officinale Flowers contain α-pinene, α-terpinene, limoene, α-thujene, β-myrcene, α-thujene and carene [352] and a research study found that anthocyanins from the Dendrobium officinale flower extended lifespan 56% and doubled the activty of superoxide dismutase levels in worms [353]. Another reason the scientists acheived such a remarkable result was because they had concentrated on creating a purified form of anthocyanins from the flower using macroporous resins, which are a type of porous microspehere.

When anthocyanins enter the body they behave much like ink or dyes used in printing or to color fabric. The power of Anthocyanins comes from their above average bioavailability. I hypothesize throughout this book that the red powder of the Philosopher's Stone is a substance with extremely high bioavailability. This causes it to enhance the bioavailability of nutrients in wine and boost its probiotics. You may have seen what happens when red wine is spilled onto a white tablecloth. It is the Anthocyanins in the red wine that sink deep down into the white linen, spoiling the tablecloth due to their deep penetration. This is why anthocyanins extend lifespan because their colored pigment allows them to be deeply carried into the tissues of the body at nano levels where they exert their antioxidant effects. This enhanced penetration is what allowed lead to be turned to gold because lead is one of the hardest substances to penetrate. So dense is lead, that it is used to

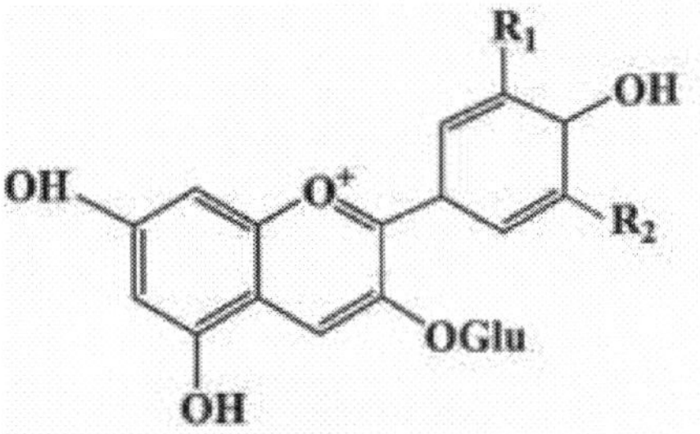

Anthocyanin

protect people from radiation and x-rays.

Anthocyanins are abundant in many berries, as well as grapes and wines. Berries are a summer food. Summer is the time of year the body undergoes significant heat stress. So nature does its part by providing a food source this time of year to help counteract this stress in the form of Anthocyanins. This is why I mention in many of my anti-aging books to eat foods that are in season. Anthocyanins are also found in abundance in cranberries, blackberries, grapes, blueberries and cherries, reaching concentrations of 3000 mg/kg fresh weight or more.

The paradox of Anthocyanins is because they penetrate deeply, they are usually destroyed by the first stage of digestion, reaching the colon partly destroyed (354) (355). Studies confirm that just 1% of anthocyanins are found in human plasma after eating foods containing anthocyanins. Plants and herbal extracts made from alcohol have their health giving substances extracted from the plant due to the low heat of the alcohol or wine via the exothermic effect. When a herb that has good longevity substances in it is made into an extract, resveratrol for example, it will never go rancid, even when kept at room temperature for months (or possibly years), compared to extracts made from other plants or substances.

From my years of personal experience making extracts (see my book titled: The Official Guidebook of How to Make Tinctures and Alchemy Spagyric Formulas) I have discovered that the spices that are almost impossible to grind up in a mortar and pestle, because they are so hard, examples being rosemary and hawthorn berry, that when they are heated, they exhibit significant increases in their antioxidant and anthocyanins levels. Raw rosemary contains between 0.2 and 1% carnosol (358) and rosemary extracts contain carnosic acid. After a period of time the carnosic acid turns into carnosol (10.3%

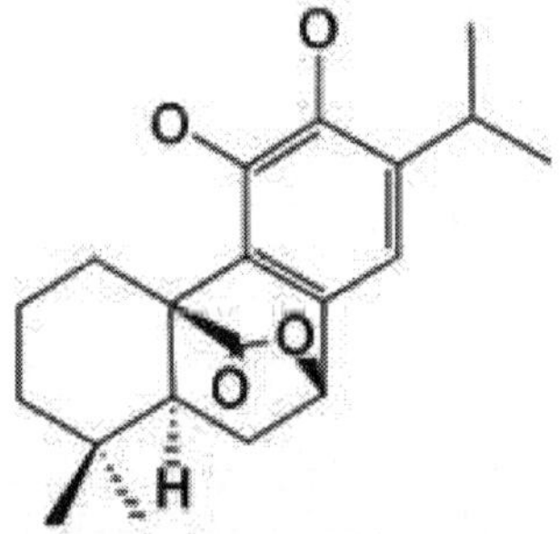

carnosol

carnosol) (358). Carnosol acts as a metal chelator for removing iron from the stomach (358). Carnosol has shown neuroprotective properties, activates NRF2, activates detoxification genes and antioxidants as well as fights tumors (358). It also detoxes the skin from air pollution (Novel Spray Dried Algae-Rosemary Particles Attenuate Pollution-Induced Skin Damage. Roberta Hoskin et al. Jun 2021). I use rosemary in the following formulas for enhancing brainpower and to increase physical endurance.

The NZT. Super Neurocognition Enhancer Formula -
6 drops of butterfly pea extract
12 drops gotu kola extract
4 drops mucuna pruriens extract
3 drops rosemary extract
3 drops cordyceps extract

The above herbs are made into separate extracts than mixed together into a final extract. Next 4 to 8 drops are added to goji berry tea.

Super Memory Formula -
2 pats ginkgo
2 parts gotu kola
1 part peppermint
1/2 part rosemary

The above is mixed amd stored in a glass jar. Next 1/3rd teaspoon is added to a cup of water. Use honey or Stevia to sweeten.

The Carnosine mix -
1 milk thistle capsule
1,000 mg of carnosine (from a capsule)
1 to 2 vitamin C capsules
Between 2 and 3 cod liver oil capsules (vitamin d)
7,000 iu of vitamin d3
200mg of grape seed extract capsules

1 teaspoon of brewer's yeast (B vitamins)
2 to 3 tablespoons of honey

The above formula works best to recover from heavy exercise.

Hawthorne Berry Heat Endurance Formula
I have used Hawthorne berry to create the "heat exertion protection formula", which uses a Maillard reaction to create the substance. The formula consists of the following -

Add the following to a pot of water BEFORE the pot boils. This is because this formula contains numerous herbs that are the hardest herbs to grind up. Hence, you are able to add these to the water before the pot boils which will bring out more of the life-giving substances. Add the following herbs -

1 tablespoon of mung beans
5 to 9 hawthorne berries (these are always hard as rock and can be bought at a natural foods store or online)
1/2 teaspoon of roobios tea
1/4 teaspoon of cat's claw
1 fingernail sized piece of Fo Ti root bark.

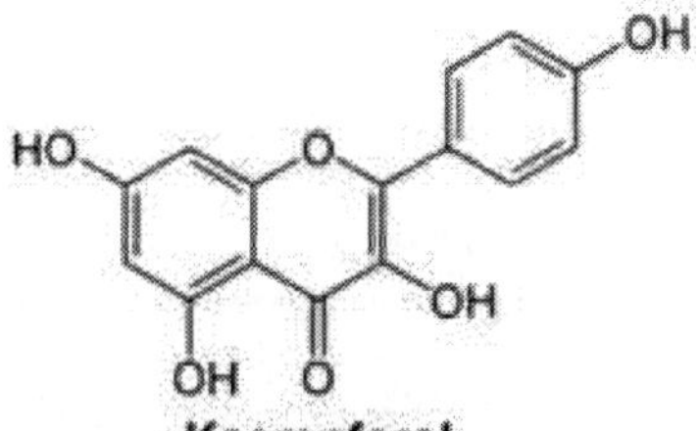

Kaempferol

When boiling is complete, strain and drink. Very, very effective for maintaining stamina in high heat environments such as desert. The main substance in hawthorn berry is quercetin. Hawthorne berries also contain kaempferol, apigenin and naringenin, with the highest levels found in the flowers (359) as well as pinene, limonene and carene (360). When Apigenin 7-O-apiosylglucoside is heated to 100 °C at a pH of 3, it transforms into apigenin 7-O-glucoside (361). However when heated at pH 5 or 7 (alkaline), it will degrade. Apigenin aglycone can also be made by combining chamomile extract with flax seed, almond or chickpea flour and incubating for 20 h at 37 °C. This causes apigenin levels to rise 4-fold (362). Similar effects are obtained mixing almond and flax seeds with celery (363). When celery leaves are heated to 100 °C for 90 min at pH 2.7, the Apigenin 7-O-apiosylglucoside in its leaves will become converted to apigenin 7-O-glucoside (364). This allows for a high aglycone flavone-rich food that can be made from heating celery leaves.

Maillard reaction Products formed by heating Apigenin

A research study (365) examined the Maillard reaction of apigenin, which is abundant in celery. This reaction also produces dimethylglyoxal apigenin (DMA), which has been shown to act as an antioxidant. The researchers of the study decided to see if dimethylglyoxal apigenin would reduce advanced glycation end products (AGE's), which cause inflammation. The researchers were astounded to find that dimethylglyoxal apigenin showed a significant effect in reducing AGE's, and stated it may reduce inflammation and act as a potent therapeutic agent for helping slow chronic diseases, helping treat chronic inflammatory disorders and that further research is necessary to examine other health benefits present in other flavonoid-Maillard reaction products that occur during thermal processing.

A recent study (366) published in June 2022, found that Apigenin in green pepper was able to strongly enhance the lifespan of worms by 23.9% and did so by reducing ROS levels (reactive oxygen species).

The Power of Polyphenols undergoing the Maillard reaction

Out of all the substances that can create beneficial compounds taking place during the Maillard reaction, Polyphenols are the very best. A research study conducted by The University of Novi Sad, Institute of Food Technology (367) set out to look at the influence of the polyphenols ferulic acid, chlorogenic acid, tannic acid, gallic and caffeic acids and how powerful they were in stopping the formation of free radicals. The researchers found that tannic acid was the most effective at inhibiting free radicals, followed by gallic, caffeic, chlorogenic and ferulic acids. In another study (368), when researchers heated Dried Blackthorn (Prunus spinosa fruits), they found **significant increases in chlorogenic acid**. There were also increases in procyanidin B2 and quercetin, and these substances also exhibited synergy with one another. One food that contains an above average number of polyphenols are apples (369).

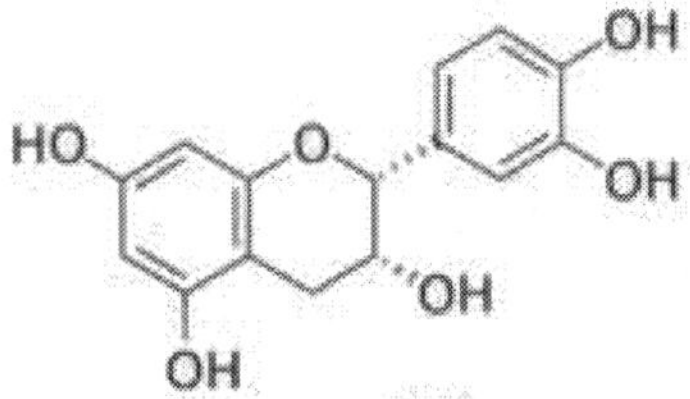

Procyanidin

For example, Joshi et al. demonstrated increases in polyphenols in Redfield apple slices that had been dried (370) and European crab apple flowers (Malus Sylvestris) have been found to have an abundance of carene, limonene, cinnamaldehyde, linalool (371), limonene, p-cymene, α-pinene, α-terpinene and camphor. Linalool and carene are the main substances found in the Earligold apple (372), which is commonly grown in Washington State USA.

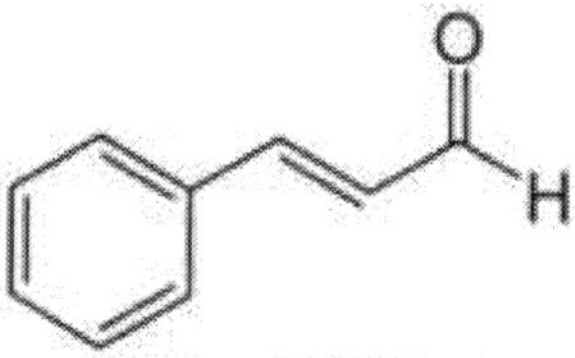

Cinnamaldehyde

Polyphenols reduce AGE's

The polyphenol flavonoids, hydroxycinnamic acids, tannins and flavan-3-ols have been found to reduce AGE's (373) (374). Hydroxycinnamic acids include caffeic, coumaric, ferulic and sinapic acids (Danijela Skroza et al. Sept 2022).

When the roots of the Lotus plant (Nelumbo nucifera) were made into a powder by first slicing the roots and then hot-air drying them at 65 °C for 12 hours and than grinding them into a fine powder (60–200 mesh) and heating them to between 30°C and 40°C, they exhibited significantly higher polysaccharides and antioxidants, displaying increases in

Lotus

their phenol levels up to 8 times (375). This could be why a hot water extract of Lotus mixed with taurine exhibits strong hepatic protective effects and increases antioxidants in obese mice (376) and fermented Lotus root has been shown to be a very powerful anti-inflammatory (377). Lotus root extract has also been shown to protect human skin (378).

The most abundant polyphenols in Lotus Root are gallic acid followed by kaemferol and quercetin (379). And what is most amazing is a thousand year old Lotus seed, which dated back to 432 BC, was able to be re-grown again. The Lotus seed was grown in the Wuhan Botanical Garden (WBG) by the Chinese Academy of Sciences in China's Hubei Province (380) (381).

In summary, one can clearly see that when the polyphenols in Lotus root are heated they exhibit a beneficial Maillard reaction that creates numerous healing and anti-aging substances. We see a similar reaction taking place in dark eggplant when it is fried. As dark eggplant is fried, it starts **showing major increases in hydroxycinnamic acids and polyphenols** (382). However if the same dark eggplant is grilled, boiled or baked, it creates the opposite effect. So the proper type of heat is key, likely reducing AGE formation. An example being that when ferulic acid is added to the baking of wheat, it reduces AGE's forming in the wheat (382). Eggplant has also been shown to contain high amounts of ferulic acid (Antioxidant Activities of Eggplant. Hanna Seprina br Sembring and Koo Bok Chin. Jul 2021.)

Tannins may be able to keep the Maillard reaction at a stable level to prevent excess AGE's from forming. For example, tannins from the Japanese Cornelian Cherry (Cornus officinalis) were shown to significantly reduce AGE's (383), and when tannins from the leaves of the Oriental Mangrove (B. gymnorhiza) were applied to Lotus roots, they showed a significant ability to reduce PPO, tyrosinase and POD, both of which are substances that contribute to the Maillard reaction. When the mangrove leaves were also applied to sweet potato, they showed a 50% reduction in PPO levels. The authors concluded this was due to the gallic acids, quercetin and coumarins present in the mangrove leaves (384). Other substances that inhibit PPO's include: cinnamon, purslane, unripe grapes, strawberry leaves and strawberry branches, coconut oil, mulberry, mango peels, pineapples, ginger, onion, chili pepper, oregano and citronella (384).

Maillard reactions of Quercetin

When red skinned ontions are fried, they show a 47.4% increase in quercetin (385). A by-product of the Maillard reaction of quercetin is phenylethenyl quercetin (8-CEPQ) which is found in beef and onion soup (386). This has been shown to stop the spread of colon cancer cells (386) and Quercetin has been shown to exert anti-inflammatory and antioxidant effects to help protect mice against atherosclerosis (387).

Apigenin

Apigenin has been shown to inhibit AGE's (388), exert anti-obesity effects and to preserve skeletal muscle in fat mice (389). Apigenin helps reduce age-related changes in heart blood vessels caused by oxidative stress. Apigenin has also been studied for its anti-cancer and anti-inflammatory properties, for its use in geroprotection and for healthy aging. Apigenin also restores age-related depletion of NAD+ levels (389). Apigenin is particually abundant in parsley, chamomile and celery.

What are NAD Levels?

NAD+ is short for Nicotinamide Adenine Dinucleotide. This substance is in every cell of your body. Its main purpose is to give your cells energy and to maintain their health. As a person ages, NAD+ levels start to decline. Methods that boost NAD+ levels are caloric restriction, fasting, sports and watching out for elevated glucose levels (390). Supplements that boost NAD+ levels include vitamin B3 (niacin, niacinamide or nicotinic acid), nicotinamide mononucleotide (NMN), nicotinic acid riboside and nicotinamide riboside (NR) (390). While NAD+ shows protective effects aginst cancer in its early stages (390), taking an NAD+ supplement during advanced stages of cancer is not advised, as it may lead to increased resistance to chemotherapy and elevated levels of inflammation (390). CoQ10 acts as a synergist for NAD+ by reducing fatigue (391) and a Maillard reaction in roasted pork called mono-MGO propyl gallate present in roasted pork has shown to exert antioxidant effects (392).

Flavone aglycones, which are abundant in celery extracts (393), are more easily absorbed by the body compared to flavone glucosides, allowing up to 10 times more luteolin or apigenin to be absorbed into the body (394). Accoding to the **USDA Database for the Flavonoid Content of Selected Foods** (395) foods that contain an abundance of above aberage levels of flavonoids are shown in the following table.

Cyanidin	Naringenin	Quercetin
saskatoon berries, broadbeans, immature seeds raw (vicia faba), cabbage red / raw (brassica oleracea (capitata group), acai fruit, apple skins, billberries, blueberries, chokeberries, raw undried cranberries, elderberries, maqui (chilean wineberry), raw black diamond plums, black raspberries, tasmanian hot peppers, purple wheat	pummelo (citrus maxima), kumquats, grapefruit	capers, coriander (cilantro), raw kale, mustard greens, raw nalta jute, peppers, ancho, roasted buckwheat

Luteolin	Apigenin	Kaempferol
mexican oregano	kumquats, raw okra, fresh parsley, raw vinespinach, (basella) (basella alba)	saffron, wild ginger
Peonidin	**Petunidin**	**High In General Nutrients**
apple skins, billberries, blueberries, chokeberries, raw undried cranberries	grapefruit, grapes, guajiru (coco-plum)	pummelo (citrus maxima), radicchio, raw (cichorium intybus) (abundant in numerous anthocyanidins), cocoa powder, black tea (very high in thearubigins)
Malvidin	**Epigallocatechin 3-Gallate And Quercetin**	**Gallocatechin**
watercress, chia seeds, cabernet franc or cabernet sauvignon or syrah or shiraz wines	green tea, brewed quingmao, oolong tea, carob flour	cowpeas, cacao beans

How Anthocyanins contribute to longevity
The term anthocyanin means flower and the term kyanos means dark blue in Greek (396). Anthocyanins are water-soluble phenolic polyphenol color pigments located in flowers, seeds, fruits and vegetables and are responsible for the colors in the fruit. The main colors being purple, red, blue and orange. This is the main reason purple corn or purple sweet potatoes are more nutrient dense compared to ordinary colored corn and potatoes.

Anthocyanin Bioavailability
Anthocyanins are very safe, with above aveage high doses of anthocyanins showing no negative side effects (397) (398). Out of the 17 anthocyanidins, only 6 become properly metabolized by the body. These are pelargonidin, cyanidin, delphinidin, petunidin and peonidin (Castaneda-Ovando et al., 2009). The following chart, taken from the **USDA Database for the Flavonoid Content of Selected Foods** (395) shows which foods contain these in

abundance.

Cyanidin	Delphinidin	Malvidin
acai fruit, raw cranberries, elderberries (chilean wineberry), raw black diamond plums (*raw brassica oleracea capitata group*), tasmanian hot peppers, purple wheat	bananas, raw billberries blueberries, grapes, black currants or juice, maqui, saskatoon berries, eggplant, black beans, cowpeas	bilberry, blueberries, jambul, grapes, grape juice, cabernet franc or cabernet sauvignon or syrah or shiraz wines, black beans, cowpeas
Pelargonidin	**Petunidin**	**Peonidin**
strawberries	bilberry, blueberries, raw grapes, raw guajiru (coco-plum), jambul, saskatoon berries	billberries, blueberries, cranberries, raw plums, saskatoon berries, cowpeas, cabernet franc or cabernet sauvignon or syrah or shiraz wines, purple wheat

Several of these annthoycians have an ability to enhance the elimination of harmful agents and toxic xenobiotic substances in the stomach, acting as a stomach detoxification mechanism (399). These anthocyanidins are malvidin-3,5-diglucoside, malvidin, petunidin, Malvidin-3-glucoside, Cyanidin 3-galactoside (*which synergizes with Acarbose*), peonidin-3-O-glucoside and Cyanidin 3-O-β-glucopyranoside (399) (which are the major anthocyanins in elderberries, blueberries, blackcurrants, black chokeberries, cowberries and whortleberries) (Dreiseitel et al., 2009).

I would like to point out that the class of anti-aging compounds known as anthocyanins, also includes cyanidin 3 glucoside, which is found in Vitis Vinifera Pinot Noir wines as well as mulberry (400). This is related to the concentrated substance known as cyanide, which if taken in excess is toxic, causes hair and teeth to fall out and in most cases death; if taken in excess. However cyanidin 3 glucoside occurs naturally in fruits and berries, especially mulberries (400). Extracts of mulberry have been found to increase mitochondrial activity (401) and mulberry leaf polyphenols when adminstered to worms in the dose of 25 mg/L, extended their life span 25% (Mulberry leaf polyphenols delay aging. Shanqing Zheng et al. Oct 2014).

As the body ages, the beneficial microflora in the stomach begins drying up. For example, research (402) has confirmed that the frailty experienced by older adults is associated with reduced populations of the good bacteria Clostridium and Bacteroidetes in their stomachs, which exist naturally at abundant levels in younger persons. Studies show (403) that anthocyanins become metabolised by microbiota bacteria, creating a drop in pH levels. This in turn produces short-chain fatty acids (SCFA's) and beneficial conditions for colonic bacteria in the lower stomach. This confirms that the good microbiota in your stomach is using anthocyanins as a rich nutritional source to produce short chain fatty acids and other beneficial substances in the body. The short chain fatty acids in your body are responsible for reducing and fighting many age related disesases. Hence they exert anticancer, hepatoprotective, neuroprotective, cardiovascular, anti-inflammatory, immunoregulatory, anti-obesity and anti-diabetes effects.

A research study (404) set out to examine the phenolic contents of numerous species of colored grains and found that black rice exhibited the highest total phenolic levels, being 86 times higher than red rice. Also the black rice contained the highest anthocyanin levels. This is common in many foods where foods that have a darker color usually contain more nutrients. This is why when fruit flies took 30 mg/d of anthocyanins from black rice (405) they experienced an extended ifespan of 14%.

Anthocyanin levels have been shown to significantly decrease during storage in coconut sugar-based jam (406). However when the jam is preserved with fructose and stevia, it helps protect the anthocyanins from decaying, probably because the fructose is undergoing a Maillard reaction. If you eat foods that contain Docosahexaenoic acid (DHA), such as fish oil, it will greatly decrase the bioavailability of anthocyanins (407).

The effect of Anthocyanins on the body after consumption

After eating foods high in anthocyanins or taking an anthocyanin supplement, they immediatly are transported to the kidneys, brain, heart and liver where they exert their antioxidant properties. In stimulated gastrointestinal digestion studies of humans, the results showed that anthocyanins remain stable in the stomach (408). However they were soon degraded in the small intestine due to the high alkaline pH levels (pH 7,15 and 16).

An in-depth research study (409) sought to look at what happens to anthocyanins in the stomach after consuming Cornelian Cherry fruit. The study found that after consuming the Cornelian Cherry fruit, that there was an increase in antioxidant levels in the body. The study also found that the anthocyanins in the Cornelian cherries were stable in the stomach, allowing them to manifest their antioxidant potential. When the Cornelian Cherry fruit reached the duodenal digestion stage, which is the stage the stomach begins to absorb nutrients, antioxidant levels in the body dramatically declined. The researchers found that Cornelian Cherry contained the anthocyanin pelargonidin-3-O-glucoside (62.99%). This was followed by the anthocyanin cyanidin-3-O-galactoside at 35.68%. When the Corneial Cherry was subjected to simulated gastric digestion, there was a slight increase in anthocyanin levels. This was also observed by Ryu et al. during the gastric digestion of Bokbunja anthocyanins, as well as Sun et al. who observed the same increase in purple rice anthocyanins, suggesting that the anthocyanins in Cornelian Cherry are highly stable in the stomach (409). I hypothesize this is due to Cornelian Cherry being abundant in polyphenols.

It is my hypothesis that foods containing an abundance of polyphenols as well as cyanidin-3-O-glycosides offer more protection against the anthocyanins during the first stages of digestion. Studies by Mazza et al. confirmed increases in antioxidant levels in studies by human participants after eating blueberries (410). This may be due to the above average levels of Cyanidin that exists in blueberries (395).

Exploring the latest methods to increase Anthocyanin stability

Research reveals (411) that the best way to stabilize anthocyanins is by co-pigmentation with other substances. This is due to the fact that co-pigments have an electron-rich structure which creates stabilization of the anthocyanins.

Pelargonidin

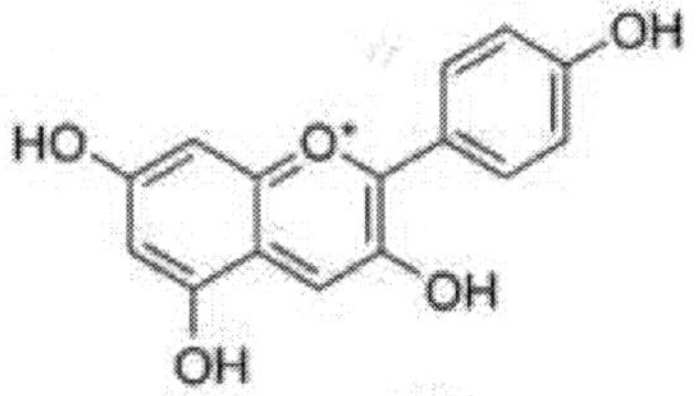

Pelargonidin

The anthocyanin Pelargonidin has an orange/red hue color and has been shown to restore antioxidant levels in the hippocampus of male rats (412). Pelargonidin is found in abundance in red radishes as well as broccoli and purple sweet potatoes (413). Pelargonidin is found in abundance in Red potatoes {Solarium tuberosum and Solarium stenotomum), which also contain coumaric acid (An abstract of The Thesis Of Luis Enrique Rodriguez-Saona. Doctor of Philisophv In Food Science And Technology. June 1998).

Stabilizing Pelargonidin

When strawberry juice was heated to 105C, the amino acid aspartic acid increased up to 33% and there was a notable difference in the stabilization of pelargonidin (414).

Goji berries contain pelargonidin (415) and have higher levels of anthocyanins when they are grown at high altitudes (415). When goji berries are heated, they show increased levels of antioxidants and carotenoids (416).

Anthocyanin Bioavailability

When anthocyanins from grape skins were added to buns and biscuits, after baking the buns and biscuits exhibited stronger protection of their anthocyanins and also showed increased bioavailability (417).

Anthocyanins improve vision

In human studies, anthocyanins have been found to enhance night vision, a person's clarity of vision and a person's ability to recognize small details with precision (418).

Black currants, which are abundant in anthocyanins, have been shown to significantly improve night vision with similar results as eating bilberries (418). Black currant anthocyanins have also been found to regenerate rhodopsin, which is a receptor in the retina of the eye (418). This has been explored in-depth in other research studies (419) 420).

Anthocyanins as a Probiotic

Berries have been found to enhance the good bacteria in the stomach (421). Studies by Cheng et al. discovered that intestinal bacteria in the human stomach feeds on substances in mulberries. This than created the following substances in the body: chlorogenic acid, cryptochlorogenic acid, caffeic acid and ferulic acid in the body (421). When chlorogenic acid reacts with glutamate, cysteine or glycine, it will produce hydroxymethylfurfural (422). Hydroxymethylfurfural is a substance that is slowly generated during storage or upon heating. Hydroxymethylfurfural is found in honey (especially when heated), UHT-milk and dried fruit. Roasted coffee can contain between 300 and 2900 mg/kg, dried plums up to 2200 mg/kg and dark beer 13.3 mg/kg. If bread is toasted, hydroxymethylfurfural levels will rise from 14.8 (at 5 minutes.) to 2024.8 mg/kg for 60 minutes.

In summary, the key is to allow the anthocyanins to reach the second phase of stomach digestion intact (colon region) so they can nourish the good microbes in your stomach, which in turn allows the body to create substances that restore the body to perfect health.

When anthocyanins are made into a juice, they are better absorbed by the body compared to a supplement. For example a rabbit study found a higher bioavailability of anthocyanins when the rabbits drank straight blackcurrant juice, compared to anthocyanins administered in a citric acid solution (423).

The anthocyanins malvidin-3-galactoside, malvidin and malvidin-3-glucoside, decrease ROS and Nitric Oxide levels and increase catalase and superoxide levels in the body (424).

The Anthocyanin Malvidin-3-glucoside

Out of the many anthocyanins, the one that is absorbed the most easily by the body is the anthocyanin Malvidin; especially malvidin-3-glucoside, which is found in abundance in Black Grape Aestivalis, Lowbush Blueberries and Lowbush Blueberry, Jam (425). Malvidin-3-glucoside is found in extremely high amounts in Vitis vinifera grape peel (426) and malvidin-3-glucoside has been found to increase Lactobacillus acidophilus and Bifidobacterium levels in the human stomach (426). Red wine contains the anthocyanin malvidin-3-glucoside (427) and a study stated that malvidin-3-glucoside in red wine may be responsible for its antioxidant and health benefits in people who enjoy drinking red wine (427). Other anthocyanins that have good bioavailability are blackberries which contain cyanidin-3-O-glucoside and delphinidin glycosides which are the main pigment in billberries (428). Malvidin-3-glucoside is also found in billberry, blueberries, High Brush blueberries, frozen blueberries, frozen blueberries, grapes, grape juice, Jambul and Saskatoon berries (429). Malvidin is found in abundance in bilberry, wild blueberries, chokeberry, saskatoon berries, Sabernet Sauvignon wine, grape juice, raw Jambul, black beans, cowpeas (blackeyes) and Açai berries (429). It is interesting to note here that a July 2023 study titled: Cabernet sauvignon dry red wine ameliorates atherosclerosis in mice, that was published by Xinlong Cheng and colleagues, found that mice that drank Cabernet sauvignon dry red wine exhibited increases in their beneficial gut microbiota, exhibited anti-atherosclerotic effects and also that the wine contained resveratrol.

Because Malvidin-3-glucoside is more abundant in red grape juice than wine, it may explain why after drinking red grape juice that Malvidin-3-glucoside levels are elevated compared to red wine (430). It is safe to say that future studies will reveal that grape juice also ameliorates atherosclerosis.

In a research study (431) involving simulated digestion of red wine, the anthocyanins peonidin 3-O-glucoside, malvidin 3-O-glucoside and cyanidin 3-O-glucoside were found to produce vanillic acid, syringic acid, p-coumaric acid (also found in rosemary), gallic acid and protocatechuic acid. Protocatechuic acid (*which synergizes with gallic acid*) has been shown to exhibit antioxidant activity (432), anti-inflammatory and anti-glycative effects (433).

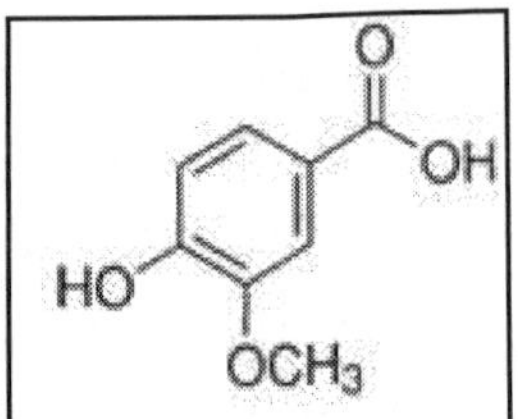

Vanillic acid

Noted health benefits of taking an anthocyanin supplement (434)

While an anthocyanin supplement is not as well absorbed into the body as the juice, you can enhance the bioavailability by taking an anthocyanin supplement than drinking or eating an anthocyanin rich fruit such as blueberries etc.

- Increased nutrients and oxygen to the eye
- Enhanced vision of patients diagnosed with open-angle glaucoma
- Improvement of eyes to dark adaptation
- A reduced risk of myocardial infarction (non-fatal)
- Improved platelet function and lipid profile
- Enhancement antioxidant capacity
- Improved insulin resistance in patients diagnosed with type 2 diabetes
- Memory protection and enhanced synaptic activity
- Neuroprotection

- Protection against neuro-inflammation and neuro-degeneration
- Promotion of healthy cholesterol levels

The berries elderberry and chokeberry also contain highly bioavailable anthocyanins (435). A study found that chokeberry was best absorbed, followed by elderberries and black currants (435). A research study (436) looked at the effects of eating 240 grams of blueberries and anthocyanin activity shortly thereafter. The study found that the participants experienced increased levels of vanillic acid, hydroxybenzoic acid, benzoic acid, caffeic acid, ferulic acid and isoferulic acids in their bodies, which remained circulating in their bodies for up to two days afterwards.

How the Stomach creates substances for good health

Fermentation by the good bacteria in your stomach has been shown to produce phenolic acid, urolithins and short chain fatty acids (437). Increased lifespan is related to healthy levels of short-chain fatty acids, and it has been proposed this has been due to the gut microbiota in the stomach producing the short-chain fatty acids (438) (439). As soon as you eat a food rich in anthocyanins, they enter your colon and begin producing the beneficial bacteria Lactobacillus/Enterococcus, Bifidobacterium, Actinobacteria, Bacteroidetes and Akkermansia, as well as reduce the harmful bacteria C. histolyticum (440). Zhou et al. found that after eating bluebrries, including taking an extract, that it exhibited prebiotic effects (441). Some bacteria in the stomach will also make anthocyanin more bioavailable by creating aglycones (441). Thus studies show that aglycones enhance the bioavailability of anthocyanins (442) (443).

Protecting anthocyanins

When blackberry anthocyanins were encapsulated with gum arabic or the oligosaccharide cyclodextrin, it created a delay in the release of anthocyanins during simulated digestion (444). Besides being an anthocyanin, Cyanidin-3-Glucoside is also a prebiotic. Cyanidin-3-Glucoside has been shown to increase the good microflora Bacteroidetes (445). Cyanidin-3-glucoside is found in black bean extract, which is available as a supplememt (446).

A study (447) found that people taking black bean extract for 2 weeks experienced no serious adverse side effects and that the cyanidin-3-glucoside was well absorbed by the stomach. Cyanidin-3-glucoside is also found in Pomegranate Juice, Blackberries, Blackcurrant, Lowbrush berries and Sweet Cherry Jam. Pomegranate juice powder has also been shown to significantly reduce edema and skin wrinkling as well as significantly increase skin moisture levels (448).

Anthocyanin levels in common foods. (per 100 gm of fresh weight). (Zamora-Ros Knaze et al. 2011)

CY = Cyanidin DE = Delphinidin MD = Malvidin
PG = Pelargonidin PO = Peonidin PE = Petunidin

Food	CY	DE	MD	PG	PO	PE	Levels
black raspberries	323			0.15	0.55		324
raw blueberries	42	92	103		23	58	320
raw chokeberries	435			1.44			437
raw elderberry	758		61	1.13			820
raw bilberries	112	161	54		51	51	430
raw red chicory	232	13					246
blackcurrant juice	16	27		1.17	0.66	3	49
raw eggplant	0.02	13		0.02			13
red raw onions	6	2		0.02	1		9
crowberry juice	16	47	61		11	26	163
raw red cabbage	72	0.10		0.02			72
raw eggplant	0.02	13		0.02			13
cranberry juice	41	7	0.31		42		91
black grape juice	1	3	58	0.02	6	2	72

The Vascular and Anti-inflammatory Activity of Cyanidin-3-Glucoside and its Metabolites in Human Vascular Endothelial Cells. Hiren P. Amin. June 2015. Norwich Medical School.

Cyanidin-3-O-glucoside has also been shown to help the body wage war against free radicals (449). Faria et al. discovered that anthocyanins peaked in the body after 24 hours of consumption and studies by Czank et al. discovered cyanidin-3-glucoside levels peaked after 48 hours, suggesting that this type of anthocyanin is a thicker anthocyanin that takes the colonic bacteria longer to digest.

Elderberries are a popular flu remedy, having been used in traditional Folk Medicine for hundreds of years. According to Phenol Explorer (**www.phenol-explorer.eu**), out of all foods, black elderberry contains significant levels of cyanidin-3-glucoside and a September 2010 study found that elderberry wine was abundant in anthocyanins and had as many antioxidants as red wine (450).

Creating a shield of protection against Anthocyanin destruction

Unlike the simple molecule diketopiperazine, which is easily absorbed by the body, only a small percentage of Anthocyanins ever reach the lower colon, allowing them to exert their anti-aging properties. Here are a few substances that have been proven to help stablize Anthocyanins during food processing –

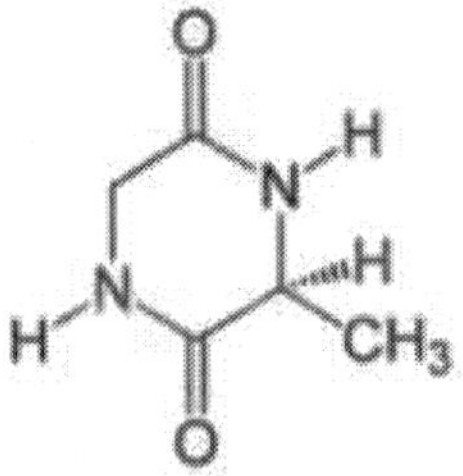

Diketopiperazine

- Xanthan gum and carboxymethyl starch has been shown to stabilize blueberry anthocyanins (451).
- Gum arabic–maltodextrin with blackberry juice and Lactobacillus acidophilus (451).
- Alginate–pectin hydrogel with blueberry anthocyanins (451).
- Encapsulating anthocyanins with beet pectin and whey protein (451).
- Encapsulating elderberry in whey proteins and pectin (451).

A homemade Anthocyanin supplement

This mixture (452) is commonly used in anthocyanin research studies. A combination of 2000 ng of Hawthorn berry powder and 67 mg of Tart Cherry powder combined with Bromelain. The bromelain was added to this formula to protect the anthocyanins from the first stages of digestion.

Chapter 13
The powerful anti-aging properties of Polyphenols

The most abundant classes of polyphenols are the tannins [453]. These larger sized polyphenols are found in leaves and bark, and protect bark against rot [453]. Polyphenols are commonly found in abundance in apples [454]. When a food abundant in polyphenols is consumed, the polyphenols are able to withstand the first stages of digestion [455] [456], thus allowing them to reach the lower colon, where they act as a prebiotic to enrich the beneficial bacteria in the lower stomach. Fleschhut et al. (2006) showed that as they entered the body orally, the anthocyanins were consumed by microflora in the intestine, where they were then made into phenolic acids [457].

Polyphenol Bioavailability

It has been estimated that between 5% and 10% of polyphenols are absorbed [459]. However 90% to 95% will accumulate in the large intestine, where they are consumed by microbiota in the gut [459]. Studies show [460] that apple polyphenols made in a smoothie from apples reached the colon, allowing for a greater bioavailability of the polyphenols compared to apples not made into a smoothie.

A research paper [460] reported that heating and grinding which occur during smoothie preparation, could increase the bioavailability of the phenolic compounds, showing that pumpkin, purple carrot and banana smoothies exhibited the most phenolic compounds. And another study [461] found that when strawberry, juçara and banana smoothie were pasteurized, they showed a higher preservation of phenolic compounds, yet showed a reduction in antioxidants.

A research study [462] that looked at 23 combinations of smoothies from fruits purchased at local Spanish markets, found increases in not only polyphenols, but also gallic acid and vitamin C levels when the combined fruits were made into a smoothie. What was most interesting in this study was that the smoothies that included apple showed the largest increases in polyphenol content. Another significant finding was that a significant correlation existed between pH level and the phenolics and flavonoids (*Phenolic acids are a major class of polyphenols*). This was due to the fact that having an acidic pH in the smoothies protected against the degradation of the phenolics. A summary of the combination of fruits made into smoothies from the study is shown below.

Increases in Nutrients of Smoothie Combinations (462)

Increases In Phenolics. Starting With Highest Combo First	Vitamin C	Gallic Acid And Vitamin C
apple, mango, banana, passion fruit	apple, pineapple, mango, carrot, coconut	oranges, mangos, and passion fruit, peach,
apple, strawberry, raspberry, blueberry, banana		apple, mango, banana, orange, passion fruit, peach, lemon
apple, blueberry, banana, pomegranate, grape, currant		
apple, pineapple, pear, kiwi, lime, spirulina		

Protocatechuic Acid	Coumaroylquinic Acid	Quercetin, Rhamnoside
banana, grape, apple, cranberry, orange, pomegranate, acai, chokeberry, lemon	apple, mango, banana, passion fruit	blueberry and pomegranate.
	apple, strawberry,	

	raspberry, blueberry, banana	
Naringin, Hesperidin	**Catechin And Hyperosides**	**Chlorogenic Acid**
oranges, passion fruit and mangos	blackberries, strawberries, grapes, currants and raspberries	berries, apples, bananas, citrus fruits and pears

When cereal is combined with polyphenols, it creates a delay in the absorption of the polyphenols. Which may be why you see cereal boxes in your local supermarket with fruit on the cereal box (463).

Polyphenols and Health

Studies show (464) that people aged 100 or over have been found to be vulnerable to the following diseases: cancer (in remission), chronic renal failure, dementia, depression/anxiety, chronic obstructive pulmonary disease, diabetes, hyperuricemia, hypothyroidism, angina pectoris, autoimmune diseases, irregular heart rhythms, osteoporosis, heart failures, hepatitis and hypertension. Many of the aforementioned conditions can be alleviated by having a healthy level of polyphenols in the diet. If you were to do an Internet search for any of conditions shown above ______ **+ polyphenols** - you would find numerous studies showing how polyphenols play a major role in keeping the body healthy and preventing many diseases. And in closing, apple polyphenols have been shown to improve the lifespan of fruit flies by 10% (465).

General Herbal Bioavailability

The plants G. uralensis (Chinese liquorice plant) and G. glabra (liquorice plant) have historically been used as bioavailability agents in many herbal formulas to increase the absorption of the main constituents in herbal formulas. These plants contain glycyrrhizin (a triterpenoid saponin). The plant G. uralensis shows up in 50% of the 283 herbal formulas mentioned in the Shén Nóng Běn Cǎo Jīng Text (*Divine Husbandman's Classic of the Materia Medica*).- Synergy in Herbal Medicines. Eric Yarnell. Dec 2015.

Chapter 14
How probiotics extend lifespan

Red wine has been scientifically proven to enhance the growth of the good bacteria in your stomach (466). These bacteria include as Lactobacillus and Prevotella (466). According to Dr. David Heber, Professor Emeritus of Medicine at UCLA Health, 70% of your immune system is located in your stomach (467). The core microflora responsible for creating healthy bacteria in your stomach are Lachnospiraceae, Bacteroidaceae and Ruminococcaeae (468). These good bacteria have been scientifically proven to create short chain fatty acids in your body, which are essential to good health. The most well-known of these short chain fatty acids are butyric acids, acetic acid and propionic acids. This was confirmed in a study by Jakobsdottir and colleagues (470) where mice fed blackcurrants made higher contents of short chain fatty acids in their bodies, and another research study (471) found that lentil proteins that were undigested / glycated, reached the colon and positively affected the stomach microbiota, causing an increase of Bacteroidetes in the colon up to 42%.

Beneficial Stomach Microflora

One of the main beneficial microfloras in your stomach is Bifidobacterium. Although there are many other strains present in your stomach, when you boil them all down, Bifidobacterium is really the main one. Keep Bifidobacterium levels healthy and all other good bacteria will also be happy. Bifidobacterium levels are increased by taking the following foods: Olive Leaf Extract, Red Wine, Red Ginseng, Grape Extract, Kudzu resistant starch, the probiotic supplement N019, fermented soymilk, rice flour enriched with resistant starch, Salacia reticulata extract, polyphenols, skim milk fermented olive leaves, alcohol free red wine, Hydroxytyrosol, Maitake extract and Cricket Chitin.

While eating crickets may seem gross to some, they are commonly ate in Thailand and Cricket powder has been shown to increase Bifidobacterium animalis up to 5.7-fold (472). Other studies (473) found that Cherries enhanced the levels of short chain fatty acids and the beneficial stomach bacteria Akkermansia when mice ate sweet cherry powder. When aged female mice were given the probiotic Akkermansia, they experienced not only an increase in their lifespan, but also were more healthy into their old age (474). Akkermansia can be bought as a supplement.

Ginseng as a Probiotic Enhancer

Extracts of ginseng have been shown to enhance the abundance of Bifidobacterium and Lactobacillus; acting as probiotic amplifiers (475). When 1% ginseng extract powder was added to fermenting yogurt, it showed decreases in moisture and increases in ash and healthy fats. It also showed stronger antioxidants (476) (477). Another research team added red ginseng extract to yogurt during its fermentation. The addition increased levels of S. thermophilus and L. acidophilus and it also showed increased antioxidants and reduced the amount of time it took the yogurt to ferment (478).

The Powerful Anti-aging Effects of Ginseng Volatile Oil

Ginseng Volatile Oil, which comes from Ginseng root, when added to worms at 25 µg/mL, caused the median lifespan of worms to live 29.26% longer than the control group and they had a survival rate of 59.33%, higher than the control group (480). ROS levels were also reduced 34.49% and SOD levels increased 27.37%. They also showed major reductions (50%) in age related pigment accumulation at 50 µg/mL than the control group. The worms also survived longer when exposed to heat shock and lived longer without any loss of reproductive capacity. This is a major finding, showing that Ginseng Volatile Oil has many powerful anti-aging properties.

Ginseng Volatile Oil Synergy

The researchers [480] studying the life extending effects of Ginseng Volatile Oil also stated that Panaxydol and Panaxynol, which are the most abundant substances in Ginseng Volatile Oil, and which are also found in Cow Parsnip (Heracleum moellendorffii) [481] may be exhibiting synergy with one another and play a large role in its significant anti-aging effects. Ginseng also contains Limonene, with the most abundant monoterpenes being α-pinene and β-pinene, found only in the American ginseng variety of plants [482].

Incorporating a diet to help boost healthy stomach microflora

The best way to boost stomach microflora is to take yogurt or kefir that contains the beneficial microflora Lactobacillus acidophilus, Bacteroides and Lactobacillus. Be sure to check the label on the yogurt as some yogurts do not include these healthy strains of bacteria. If you want to get the most out of the probiotic, add a prebiotic to the yogurt and stir it in before eating. I have a found a combination of coriander and astragalus powders in a 1:1 ratio with 0.5 ratio of olive leaf powder. This combo is mixed together and 1/2 a teaspoon of this is mixed in with the yogurt with 1/2 teaspoon of cinnamon. You can really feel the difference after eating yogurt when adding this combination, especially when you leave the mixture in the yogurt for 10 to 20 minutes before eating it.

Blackcurrants, blueberries, pomegranate, strawberries, jussara fruit and tart cherries have also been documented to increase Bifidobacterium levels in the stomach [483]. Below are some additional foods that enhance healthy micoflora [484].

- 2.35% freeze dried strawberry taken as a supplement for 10 weeks increases Bacteroides in mice.
- 250 mL of wild blueberry consumed for 6 weeks increases Bifidobacteria levels.
- A high fat diet with pomegranate peel extract (taken

at 0.2% in water) will significantly increase Bifidobacterium and Bacteroides-Prevotella spp. levels.

- Black raspberry at 10% will increase Bacteroidetes levels.

- Drinking 8 ounces of tart cherry juice for a minimum of 5 days will increase Bacteroides and Lactobacillus levels.

- A standard diet that includes freeze dried raspberry, blackcurrant or blackberry significantly increases the levels of short chain fatty acids in the body.

- Taking a Cranberry extract will increase Akkermansia levels up to 30%.

Anti-aging secrets of the Grape Seed

Grape seed happens to contain an abundance of gallic acid, protocatechuic acid (485) as well as proanthocyanidin (PA) (486). Proanthocyanidin strengthens collagen-based tissues due to its ability to enhance collagen cross-links. As temperature is increased in grape seed extract, it begins showing increases in gallic acid and the procyanidins B1 and B2 (486).

Using Grape Seed Extract for Tooth Remineralization

Tooth cavities start to increase as one grows older, being especially common among elderly people. A report published by the National Institute of Dental and Carniofacial Research titled: Dental Caries (Tooth Decay) in Seniors (Aged 65 and older), stated that 19.5% of people aged 75 years or older exhibited tooth decay, with whites and people with above average levels of education and lower income showed higher rates of tooth decay.

A research study (487) looked at using grape seed extract that contained 97.8% proanthocyanidins to determine how effective it was at repairing the collagen rich root tissue of human teeth; helping to remineralize teeth. The researchers found that grape seed extract was able to assist in the remineralization of teeth with cavities and did so similar to that of Fluoride, but through a different mechanism. The authors of the study concluded that grape seed extract be recommended as a promising natural substance for non-invasive methods in treating cavities.

A clinical trial (488) involving 38 diabetic patients were fed a high-fructose diet and given 2 grams a day of grape polyphenols. The study found that the grape phenolics protected the diabetics from protein carbonylation, fructose-induced oxidative stress and lipid peroxidation. It also helped with insulin resistance. Another study (489) involved 37 adults that were given quercetin 3-glucoside (160 mg/day) and epicatechin (100 mg/day), both of which are abundant in grapes. This researchers found that the quercetin, but not the epicatechin, decreased their plasma methylglyoxal concentrations. Increases in methylglyoxal levels are associated with cardiovascular disease and increased mortality in individuals diagnosed with Type 2 Diabetes. The authors of the study concluded that Quercetin may be a good treatment strategy for diseases where methylglyoxal levels exist in the body at above average levels. The authors of the study (489) concluded that grape phenolics play a good role in helping the body defend itself against protein glycation.

Reducing AGE's in food using Grape Seed Extract and Grape Pomace Phenolics

When the polyphenols epicatechin, chlorogenic acid, rosmarinic acid, naringenin and quercetin were added to cookies, quercetin was the most effective at inhibiting advanced glycation endproducts (AGE's) in the cookies (490). When ferulic acid, caffeic acid, quercetin and gallic acid were tested in bread, they were all found to reduce AGE's, even at low concentrations,

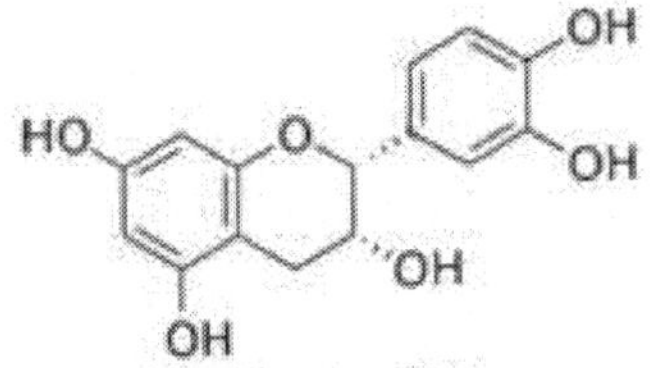

Epicatechin

with catechin being the most efficient (491). When grape phenolics were added to baked bread, the losses of catechin, quercetin, gallic acid, ferulic acid and caffeic acid were found to be reduced by 51%, 49%, 86%, 75% and 47%, respectively (492).

Other studies (493) have found that balady bread (*bread that is rolled flat and baked*) that had grape seed powder added to it contained high levels of antioxidants, polyphenols and flavonoids after baking. The best results were obtained when the flour contained 10% grape seed extract. Phenolic extracts from white or red grape skins show stronger anti-glycation effects than those from commercial nutraceutical preparations (494). The following substances show good anti-glycation activity, from best to last in order of effectiveness: >>>>>> quercetin 3-O glucoside >>>>>> malvidin 3-O glucoside >>>> catechin >>> procyanidin A2 (494). Also the substances sage, cinnamon and rosemary are also effective anti glycation substances.

Grape seed extract has been shown to increase fermentation time in yogurt, as well as create a yogurt with a softer texture (496). It also did not decrease L. acidophilus levels when added at concentrations of 1.5 and 3.0 g L. However it did decrease L. acidophilus levels when added in lower concentrations, suggesting that adding grape seed extract to yogurt may exert antimicrobial effects (497). Grape seed extract also extends the shelf life of yogurt (498) and the addition of grape pomace powder to yogurt improved phenolic levels (Demirkol and Tarakci 2018). The addition of organic pomegranate peel powder added to yogurt at just 2% enhances flavonoid and phenolic levels in the yogurt by 97.72% and 96.52% respectively (Hamid et al. 2022) (499). Adding blackstrap or carob molasses to yogurt (500) enhances not only the stir texture, but also increases the solids, fats, proteins and ash from 15.57% to 24.05%, 4.29% to 3.37%, 5.33% to 6.27%, and 0.80% to 0.88%, respectively (Shalabi 2022) and adding 0.5% grape seed extract to buffalo stirred-type yogurt (501) will show significant increases in levels of the 2,2-diphenyl-1-picrylhydrazyl radical, increases in antioxidants and it also reduces the damage caused by free radicals. DPPH is a method used for assessing antioxidant activity in foods such as fruits, wine, herbal teas, yogurt etc.

Exploring the natural Antioxidant power of Chamomile

Chamomile (Matricaria chamomilla) is a plant species in the Asteraceae variety. Extracts from this plant have shown it can help preserve yogurt and cottage cheese [502]. This most likely is due to the large number of polyphenols and the flavonoid glycosides luteolin, luteolin, apigenin and apigenin-7-O-glucoside which are particularly abundant in chamomile [503]. This is why when chamomile is freeze-dried or oven-dried, it shows increases in its polyphenol levels [504]. Indeed studies [505] show that increased polyphenols contribute to enhanced antioxidant levels. Polyphenols in chamomile vary by country. For example, chamomile purchased from a local market in the United Arab Emirates contained more polyphenols than chamomile found in Egypt. The highest polyphenol contents were from chamomile grown in Italy [506].

Yogurt of the stirred-type yogurt that has had grape seed added shows good effects in reducing the pathogens S. aureus (*which causes tooth decay*) and E. coli (Tami et al. 2022) [507]. When stirred-type yogurt has 15% carob molasses added to it, it exhibits potent antimicrobial activity against S. aureus, Enterococcus faecalis, Shigella spp., Bacillus cereus, E. coli and Salmonella typhi, compared to control samples which showed zero inhibitory effects (Shalabi 2022). When I wrote my book titled: The Complete Guide to Natural Toothache Remedies and Re-mineralization, my research revealed that blackstrap molasses was one of the few substances that had a perfect balance of calcium, and phosphate necessary for not just keeping teeth healthy but for natural tooth remineralization. And in closing, adding just 0.5% grape seed extract to yogurt significantly enhances the destruction of cancer cells. (Tami et al. 2022) [509].

Chapter 15
Exploring anti-aging substances produced via the Maillard reaction

Ancient alchemical texts state that the original red powder was made by exposing a substance over low heat for a long period of time, which gradually allowed it to experience a series of Alchemical changes. The substance was than dried out and made into a powder. The red powder made one immortal by adding just a few grains of it to red wine. If too much was added to the red wine, the person's hair, teeth and nails would fall out, which would than gradually regrow again over the period of a few weeks.

Whether or not you believe in the story of the Philosopher's Stone, the scientific evidence shows some interesting reactions when amino acids are heated. These substances, a by-product of the heat process, reduce the damage of free radicals, which are the cause of all aging. When lead is transmuted into gold in the Philosopher's stone, there is a Maillard reaction taking place where the red powder is first encased in a ball of wax the size of a grain of rice, than added to molten lead which turned the lead into ultrapure gold. This means substances that create Maillard reactions are smilar in strudture to the red powder of the Philosopher's Stone. This adds strength to my hypothesis that the alchemical red powder enhances the bioavailability of the red wine because the making of the red powder is similar to the Maillard reaction. It is a substance that enhances bioavailability of the health giving nutrients in red wine.

A similar effect to that of wrapping the red powder in a ball of wax was mentioned in a July 2019 research study titled: Effect of using cinnamon oil encapsulated in maltodextrin as exogenous flavoring on flavor quality and stability of biscuits, that was published by Hoda H. M. Fadel and colleagues. The researchers wrapped cinnamon essential oil in maltodextrin and then baked a batch of biscuits. The researchers were stunned to discover higher levels of caryophyllene, cinnamaldehyde and eugenol in the finished biscuits ($P < 0.05$). The authors in the study stated this was the result of the Maillard reaction occuring as the biscuts were baked. Research is starting to show that the Maillard reaction has played a major role in determining longevity in mammals. For example, the Maillard reaction is responsible for genetic regulation homeostasis activity in mammals. For example, Maillard reactions involving Lysine and Glucose produce substances that enhance stress tolerance and extend lifespan in worms (511). The research (511) also states the Lysine / Glucose Maillard reaction creates declines in motility by delaying aging and that out of all the amino acids tested, lysine extends the lifespan of worms via upregulating NRF2 (Edwards et al., 2015). What is interesting is this study found the effects were dose dependant. This means that only a small amount was necessary to get the best results.

What is the Maillard reaction?

The Maillard reaction is a process where certain substances are generally heated between 280°F and 330°F. This reaction creates new substances that act as potent free radical fighters. The Maillard reaction is really a form of controlled Age Glycation (advanced glycation end products), whereby the proper temperature and length of heat are applied to create beneficial substances. When the Maillard reaction occurs in food high in animal protein, it can create AGE's. This reaction usually takes place at extremely high temperatures because the amino acids turn rancid. You can take foods to offset AGE's such as lemon and limes because they help make the proteins in the foods more digestible. An example is adding lemon juice to oatmeal which makes the protein easier to digest by the body.

A Maillard reaction that is beneficial for the body can occur in the presence of the amino acids lysine and arginine. Coffee that has not been lightly roasted contains arginine. The arginine is reduced after 4 mins of roasting at 240 degrees Celsius (512). When coffee beans are roasted they exert significantly higher antioxidants than green coffee beans. The study (512) found that coffee's antioxidant power is influenced by where the coffee comes from, if it is organic or conventional, the roasting level of the beans and duration of infusion. Lightly roasted coffee contains high polyphenol levels with longer roasting times reducing polyphenol levels. However antioxidant levels in coffee begin to rise as polyphenols levels start to decrease (513). I would like to point out here that human studies have found that coffee has been shown to extend lifespan and that Zaro Aga, who was reported to have lived to 170 years of age, worked as a coffee brewer for the local city council in his hometown and also enjoyed drinking wine (514) (515). Zaro was born in 1774 and fathered 25 sons, eight daughters and had been married 11 times. Two of his sons were documented to have lived to over 100 years of age (514) (515). Could it be that the substances from the coffee were responsible for his long life? Perhaps other dietary factors may also have played a role in his long life.

Examples of the Maillard reaction and the formation of longevity substances

Thyme. When heated to 180°C, thyme exhibits significantly higher free radical-scavenging activity (516).

Onions. When onions are fried, they show a 47.4% increase in quercetin (516).

When grape seed extract has been added to chicken at 0.5 g/kg before frying, it significantly reduces AGE's ($P < 0.05$). However when excessive grape seed extract was added, (0.8 and 1 g/kg) it actually caused more AGE's to be created. Air frying the chicken gave significantly better results than the deep frying method ($P < 0.05$) (517).

When amino acids are combined with starches, it has been shown to improve the nutritional value of the food. For example, when the amino acids lysine and aspartic acid are added to sweet potato starch, it improves its nutritional value (518). The amino acids lysine and arginine have positive charges when at a neutral pH and aspartate and glutamate, are more acidic amino acids, having negative charges a neutral pH.

If you have ever taken out 1/3rd of the powder from a resveratrol capsule and made it into a tea, it has an ashy burnt odor, like it is a by-product of the Maillard reaction. From my years of anti-aging research foods that are high in ash content are usually some of the best anti-aging substances. Heating a combination of Peony seed, chicken fat, and cysteine or heating sesame seeds with cysteine and xylose, thiamine or methionine produces the substance pyridine (519) and pyridine is produced when coffee is roasted (520). Pyridine possess antixodant properties and extends the life span of worms by reducing heat induced oxidative stress, when adminstereed in dose-dependent amounts (521). It has also been found to significantly reduce advanced glycation end products (AGE's) (521). Maillard reactions will cause food to change color, alter proteins and affect the digestion of proteins (522). The Maillard reaction can be created in the lab by heating aspartic acid, lysine, arginine, tyrosine, serine or cysteine with the sugar sucrose (523).

Summary

It is the heating of sugars (*glucose for example*) with amino acids (*lysine for example*) that creates the Maillard reaction. This is because the sugar degrades when it becomes heated, acting as a catalyst, which than leads to anti-aging substances being produced as a by-product of the reaction (524).

Besides sugars evaporating in the presence of heat (between 20 and 55%), certain amino acids will also undergo a decrease up to 25% depending upon the chemical constituents present in the formulation. Under certain conditions, when excess sugar is added, the amino acids will evaporate completely (524).

Chlorogenic acid and the Maillard reaction

One main substance that comes from the Maillard reaction involving coffee is chlorogenic acid, with the highest levels found when the coffee beans are lightly roasted (525). When chlorogenic acid is combined with the amino acid Leucine, it exhibits strong synergistic effects, reducing fatty acid accumulation in muscle cells up to 76% (526). The total chlorogenic acid contents in coffee ranges from 187.7 to 295.6 mg/100 mL when the coffee has been made from lightly roasted coffee. These levels vary between 24.2 and 41.3 mg/100 mL in dark roasted coffee beans (527). Studies by Tfouni et al. demonstrated that brewing coffee in boiling water without a filter yielded higher levels of chlorogenic acid, compared to filtering the coffee (528). Chlorogenic acid levels are also found in higher levels in espresso coffee brew. Besides caffeine, coffee also contains ferulic and coumaric acids (528).

Apple seeds contain an abundance of chlorogenic acid, with the apple core having the highest amounts (529). The herbs chrysanthemum, rtemisia annua, kuding tea, honeysuckle flower and purple sweet potato stem also contain abundant levels of chlorogenic acid (529).

Chlorogenic Acid and Lifespan

Chlorogenic acid has been shown to extend lifespan in mice up to 24% (530) and when mice were given a plant extract of chlorogenic acid, it up-regulated expressions of their SOX9 gene, assisting in the development of their cartilage (531). Chlorogenic acid acts as a probiotic, escaping destruction from first stage digestion. As it reaches your colon, the good bacteria in your stomach uses the chlorogenic acid as a food source to produce several metabolites (532) such as coumaric acid and benzoic acids. An oral supplement of TAC-101, which is a derivative of benzoic acid (534), has been shown to extend the **lifespan of mice with cancer up to 142%**, being more effective then the prescription anti-cancer drugs CDDP,

CPT-11 and 5-FU (535).

Pyrrole

Heating coffee to the right temperature and duration produces the nitrogen-containing heterocyclic substance called pyrrole. Pyrrole is the source from which many medical drugs are derived. It is found in vitamin B12 and the yellow bile pigment bilirubin and the green substance biliverdin (536). Many anti-aging substances that are highly bioavailable are pigment related because the pigment has very small molecules which are easily absorbed by the body. Anthocyanins are one example. Pyrrole is also used as an antibiotic, a fungacide, as an anti-inflammatory, to reduce cholesterol and for treating tumors (537).

The Maillard reaction in anti-aging foods

I hypothesize that when resveratrol is exposed to heat, it undergoes a Maillard reaction which produces beneficial substances. I have stated throughout this book that boiling resveratrol from a reseveratrol capsule greatly enhances its effects. I have found that doing this enhances memory. This should not be donex more than 3 days in a row, as resveratrol has accumulative effects and having an excessive buildup of anti-aging stubstances in the body is not always healthy unless you are doing intenstive exercise. It is best to take a 3 to 4 day break than start the resveratrol tea again. I have also found resveratrol to be of great benefit as an exercise recovery supplement. Adding xylitol, allulose, honey or stevia to sweeten the tea is highly recommended because Resveratrol tastes bitter as most anti-aging foods taste bitter. The enhancement in memory recall could be because resveratrol has shown protective effects against neurodegenerative diseases and because it enhances the natural secretion of neurotransmitters in the body (538). Excessive resveratrol in the body has been shown to negatively affect enzyme activity in the stomach and brain (539).

Maillard reaction type effects can also happen when foods are subjected to sudden stress. For example, when carrots or spinach are cooked or pureed, carotene levels increase 3-fold, compared to eating them raw [540]. The effects produced by the Maillard reaction are similar to biostimulants (plant growth regulators), which have been shown to increase the polyphenols in plants and increases in polyphenols also take place during food processing.

Several recent studies [541] have reported that substances may show increases in antioxidants and exhibit enhanced bioavailability after heating during food processing. (Al-juhaimi et al., 2018, Barba et al., 2017, Castro et al., 2020, Ribas-Agustí et al., 2018). For example the best known one is tomato sauce, which shows increased lycopene levels. Other examples include increased levels of polyphenols in potatoes, increased levels of anthocyanins in pasteurized strawberries, increased levels of phenolics in eggplants, increased levels of hydroxycinnamic acid in Maqui fruit, increased levels of ferulic acid, epicatechin, ellagic acid, gallic acid, rutin and epigallocatechin in aprocits, increased levels of polyphenols in black currants, redcurrants, raspberry and boysenberries, increased levels of phenolics and anthocyanins in grape juice and increased levels of flavonoids, phenolics, gallic acid, gallocatechins and procyanidins in Lychee [541].

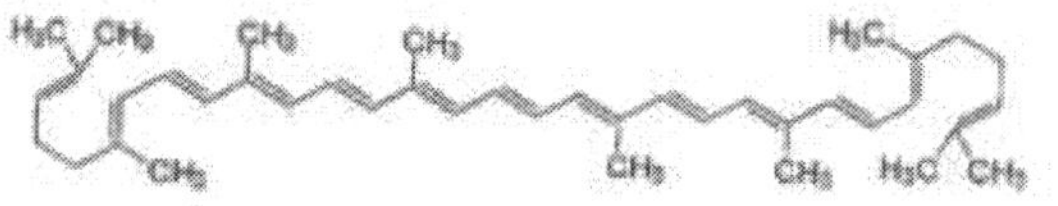

Lycopene

When Lye is slowly mixed in with water, rasing the pH of the water to a more alkaline level, the life-giving elements known as Ormus fall out of the water. If you have ever tried this, the Lye undergoes a very hot chemical reaction when mixed with water.

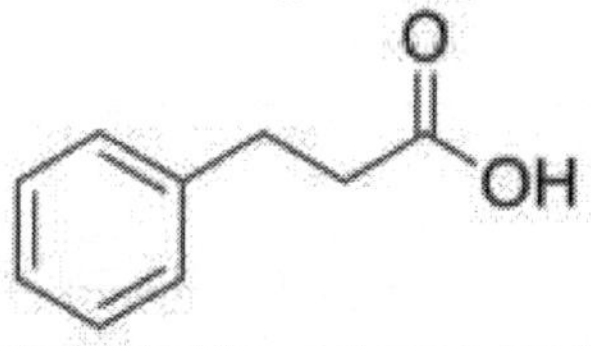

Hydrocinnamic acid

Maillard Reactions take place in -

- the making of Ormus (Exothermic reaction via Lye & water)
- exercise (internal body heat)

- making extracts with alcohol (Exothermic reaction via alcohol)
- a slight body fever (boosts immune system (542))

Beneficial Maillard reactions occurring in Coffee

Coffee has been shown to extend the lifespan of worms up to 52% (543) (544 (545) and when medium coffee is lightly roasted it produces cyclodipeptides (546). But that is not the only beneficial chemical reaction that takes place, it also produces salicylic acid (547). Above average levels of salicylic acid are also found in cumin (*Lagoecia cuminoides*) (1474.7 mg/100 g) (548). The predominant substances in cumin are chlorogenic acid, rosmarinic acid, vanillin, hesperetin and hesperidin (549). Salicylic acid is also found in turmeric and paprika (548). Coffee peels contain not only chlorogenic acid, but also salicylic acid and vanillin (551).

When worms were given salicylic acid, they lived 19% longer than the control group as well as lived healthier and happier lives (552). The animals were also more fit even as they got older. When sows had salicylic acid added to their diet, the piglets showed an improved survival rate, healthier levels of iron in their body and also had healthier birth weights compared to the sows that were fed a standard diet (553). Drinking cranberry juice naturally raises the levels of salicylic acid in your body (554).

Salicylic acid is used by coffee plants to protect themselves against fungus. Organic coffee contains significantly more salicylic acid ($p < 0.0001$), compared to non-organic coffee (555). Coffee also contains quercetin-3-O-glucoside (556) and quercetin-3-O-glucoside which has been shown to extend the lifespan of worms up to 39% (557).

When quercetin-3-O-glycosides is heated, it shows increases in its antioxidants (558). Quercetin-3-O-glycosides are found in abundance in red raspberry jam, bottled tea, raw broccoli, Blackcurrants, Black chokeberry (highest amounts), Greencurrant and Rowanberry juice (559).

Foods that show increases in antioxidants when heated -
In these foods, scientists hypothesize that increased polyphenol levels in heated foods may be the result of a possible breakdown of the tannins present in the foods as they are heated to the right temperature (*the temperature sweet spot*). This chemical reaction leads to increases in antioxidants and other favorable substances (564).

Antioxidant Levels increase when these are boiled (560)
Kenikir leaves (54.93%), Papaya leaves (59.04%) and Turi flower (53.84%).

A brief submersion in boiling water enhances the antioxidants of the following (561)

- Bean sprouts briefly submerged in boiling water show increased antioxidant levels up to 49.27%.
- Thai basil leaf. Boiling between 1 and 5 mins.
- Sweet potato leaf. Boiling between 1 and 5 mins.
- Cooked artichokes exhibited a significant increase in antioxidants after steaming (up to 15-fold) and the most abundant amino acids in artichokes are leucine and lysine (562).

- Boiled or steamed Pumpkin increases Vitamin A levels (563).

- Leaves of the Ceylon cow plant (G. lactiferum) when boiled, exhibit a 167% increase in polyphenols (564).

- When broccoli is cooked it results in significantly higher levels of tocopherols, lutein and β-carotene (565).

- Mallow, perilla leaf crown daisy, spinach, broccoli, chard and zucchini, when cooked, show increases in Tocopherol (vitamin E). However carrots, potatoes and sweet potatoes show significant decreases in

Vitamin E when cooked (566).

- When carrot is made into a puree it doubles its carotene bioavailability, compared to boiled-mashed carrots (567).

- Apigenin-7-O-glucoside from Chrysanthemum increases its effects when heated to a temperature of between 50 and 60°C. Any temperature above this causes a decrease in the effectiveness of apigenin-7-O-glucoside (568).

Onions

When green onions are fried in soybean oil, they show significant increases in limonene, linalool, unsaturated fatty acids, monounsaturated fats and polyunsaturated fats (healthy acids and fats) (569).

When green onions are fried in olive oil, they show a significant increase in unsaturated fatty acids (healthy acids) and monounsaturated fats and polyunsaturated fats (good fats) (569).

Tomatoes

When tomatoes are cooked, it results in a significant increase in the bioavailability of both the naringenin and the chlorogenic acis (570). The bioavailability of lycopene increases when olive oil is added to diced tomatoes (571).

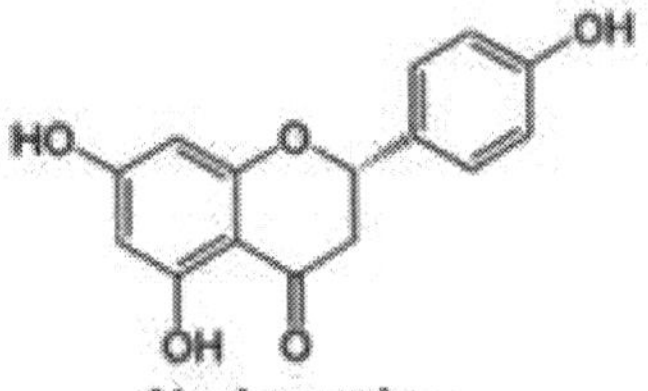

Naringenin

Cooking tomatoes in olive oil at 100 °C increases lycopene levels, with even better results using Pulsed Electric Fields (PEF) (572).

Tomato paste has between 22% and 380% more lycopene compared to lycopene in fresh tomatoes (Gärtner et al., 1997, Porrini et al.,

1998) and heat-processed tomato juice contains higher lycopene levels compared to unprocessed tomato juice. Scientists state (572) that the enhanced bioavailability of lycopene comes from the cells in the plants becoming disrupted during the thermal and mechanical processing (Gärtner et al., 1997, Stahl and Sies, 1992).

Using Ultrasonication (40 kHz, 250 W, 4 °C) for 20 minutes on lettuce, green zucchini, tomato and red pepper increases bioavailability up to 150% (Lafarga, Rodríguez-Roque, Bobo, Villaró, & Aguiló-Aguayo, 2019) (573). Ultrasonification also enhances the germination of seeds and improves crops (574).

When black plums are dried, their bioavailability increases up to 250% (575).

Drying apple skins in heated air enhances the bioavailability of the polyphenols, followed by freeze drying **(576)**.

Chapter 16
Exploring anti-glycation end products (AGE's) in detail

Advanced glycation end products (AGEs) are molecules created by the classical Maillard reaction. AGE's occur most often when foods, especially proteins such as lysine, are heated above 240 C. They were discovered in 1912 by scientist Louis-Camille Maillard.

In 1981 Dr. Monnier and Dr. Cerami published their groundbreaking research paper titled: Nonenzymatic browning in vivo: possible process for aging of long-lived proteins, which hypothesized that Maillard reactions are intimately involved in the aging process [577]. Their original hypothesis was that over long periods of time, an accumulation of AGE's contributed to aging (Bjorksten, 1968; Monnier, 1989; Monnier et al., 1988; Sell and Monnier, 1989). They also stated the gradual buildup of AGE's altered the function and structure of proteins (Gugliucci and Menini, 2017; López-Otín et al., 2013). It is now a fact that AGE's contribute to many age related diseases such as atherosclerosis, diabetes, oxidative stress, inflammation and many neurodegenerative diseases [578].

As proteins react with glucose and fructose, both in the body and in foods, they contribute to AGE's, which causes protein glycation to take place in the body's tissues and fluids [579]. The slow gradual build-up of AGE's is accumulative, taking decades to reach peak levels in humans. Research over the past few decades has shown that there is a gradual age-dependent increase in cross-linking and browning (*a negative Maillard reaction*) as AGE's gradually build-up over time in collagen [580] (Monnier et al., 1984; Monnier VM, Stevens VJ, 1981).

Levels of AGE's are higher in older populations

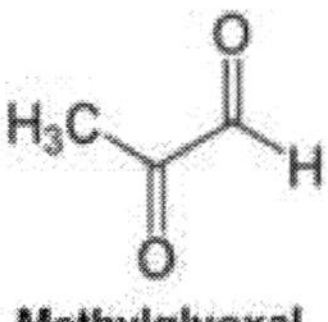

Methylglyoxal

One AGE known as carboxymethyl-lysine, is positively correlated with an increased risk of cardiovascular disease (CVD) or general-cause mortality death in older adults (Semba et al., 2009) (581) and a specific type of AGE called Methylglyoxal has been shown to accelerate vertebral bone loss in aged diabetic mice (582).

Most AGE's escape digestion and are delivered to the colon where they wreck havoc on the good bacteria in your lower stomach. When meat is grilled, it leads to a significant increase in AGE's, especially in animal-derived foods rich in fats. However if these foods are made into a stew, boiled or poached, it decreases AGE formation (583). If you cook pork in citric acid (*which is an acidic substance with a low pH*) AGE's will be reduced up to 54% and cooking it in acetic acid reduces AGE's up to 48% (583).

Studies now show (584) that the highest levels of AGE's occur in grains or nuts which have been processed at high heat in order to preserve them, including canned meats. Foods that contain the lowest numbers of AGE's are vegetables, fruits and coffee (584). AGE's don't just occur in overcooked foods, they also occur in processed foods that are kept at low temperatures (584). This includes butter, cream cheese, high-fat spreads, mayonnaise and margarine. If milk products contain above levels of water moisture, such as yogurt, ice cream or pudding, they usually have lower AGE's (Uribarri et al. 2010) (584).

There are some good online databases showing AGE levels in numerous food groups. These can be downloaded at your convenience. Interestingly, broiled Tofu contains extremely high levels of AGE's compared to soft raw tofu, which has some of the lowest AGE's out of all the foods.

Online AGE Database Resource #1

Advanced glycation end products in foods and a practical guide to their reduction in the diet. Uribarri J., Woodruff S., Goodman S.,

Cai W., Chen X., Pyzik R., Yong A., Striker G.E., Vlassara H. J. Am. Diet Assoc. 2010;110:911–916. doi: 10.1016/j.jada.2010.03.018.

Online AGE Database Resource #2
Analysis of advanced glycation end products in selected food items by ultra-performance liquid chromatography tandem mass spectrometry: Presentation of a dietary AGE database. Scheijen J.L.J.M., Clevers E., Engelen L., Dagnelie P.C., Brouns F., Stehouwer C.D.A., Schalkwijk C.G. Food Chem. 2016;1:1145–1150. doi: 10.1016/j.foodchem.2015.06.049.

Research studies (586) have sought to determine if restricting AGE's in the diets of mice throughout the course of their life time extends their lifespan. Such a study was performed by (Cai et al. 2007) which found that the mice lived longer when they had a reduced number of AGE's in their diet. The authors in the study concluded that restricting AGE's in the diet may be helpful in the prevention of age associated various disorders.

When foods are heated at low temperatures and with amino acids and certain sugars, it will act as a type of antidote to AGE's. For example Chlorogenic acid, a substance produced by lightly roasting coffee (587), has been shown to inhibit the formation of AGE's. As a matter of fact Chlorogenic acid was so effective at inhibiting AGE formation, that it was more effective than the pharmaceutical aminoguanidine, which is commonly used to inhibit AGE's (588). Another substance that inhibits AGE formation, also produced in significant amounts when coffee is heated to high temperatures, is salicylic acid (aspirin) (589).

Aspirin works to inhibit glycation by helping to liberate the amino acids in proteins (590). This than blocks the ability of sugars to remain attached to the proteins (Caballero et al. 2000; Malik and Meek 1994) which create the beginning stages Age Glycation.

Infant formulas bought in the supermarket have been found to have a 70-fold higher amount of AGE's (CML) compared to AGE's present in natural human breast milk (Sebekova et al. 2008)

[591] and the heating of milk greatly increases its AGE content [592].

Methods to reduce AGE's during high temperature cooking [593] -

- The addition of amaranth flour or amaranth protein isolate to cookies, fried tortilla chips or baked tortilla chips.
- The addition of spices to buckwheat ginger cakes.
- The addition of glycine or glutamine to potato crisps or french fries.
- Soaking potato chips in taurine.
- Soaking potato strips in a solution containing vitamins.
- Fried potatoes - Immersing potatoes in solutions containing natural extracts.
- Roasting coffee under a vacuum.
- The addition of green tea flavonoids to UTH milk (ultra heat treatment).
- The addition of glutathione to grape juice.
- Addition of salts, amino acids, antioxidants to ripe olives.
- The addition of olive oil and oregano phenolic compounds to fresh potatoes.
- Adding resveratrol to yogurt or baked milk reduces AGE's [594]. This was shown to reduce furosine up to 60%. It also completely inhibited the formation of

pyrraline (595), which has been implicated in diabetes and other related complications

Foods cooked at lower temperatures or ate raw are relatively low in AGEs (596). However the problem with this is that eating some foods raw reduces their nutrition value because the nutrients are not as well absorbed into the body. For example, people on a strict raw foods diet have lower levels of healthy cholesterol, vitamins B12 and D3 and lower amounts of triglycerides and zinc in their bodies (597). Raw foodists enjoy better mental health, with apples and raw bananas having the most benefit (598). The key is to find which foods are low in AGE's and when eaten raw, have high nutrient value.

One of of the main substances in black cumin seed (Nigella sativa), **Thymoquinone**, has been shown to reduce the early phases of AGE's (599) and when black cumin seed is fed to aged mice, it has been shown to help protect their livers and brains (599). Other substances proven to reduce AGE's are Vitamins E, B6 and B1, the Microalgal extract Chlorella, rutin from tomatoes and epigallocatechin gallate, found in abundance in green tea (601).

Rutin has been shown to exhibit significant effects to stop AGE's that caused the narrowing of joint spaces and the calcification of cartilage tissues in mice diagnosed with osteoarthritis (602).

Creating the proper pH level in foods that are naturally high in AGE's can also help reduce increases in AGE's during cooking. For example, when beef is marinated for an hour in vinegar or lemon juice, it will show half as many AGEs when it is cooked (Uribarri et al. 2010) (603). Green tea has been found to inhibit AGE formation in the collagen of diabetic and aged mice, as well as in vitro bovine serum albumin (Babu et al. 2008; Nakagawa et al. 2002; Rutter et al. 2003) (603).

Yerba Mate

Herbs that reduce the accumulation of AGE's

When it comes to herbs and herbal infusions, Yerba Mate (Ilex paraguariensis), Guava (Psidium guajava L.), Sword-leaf Dogbane (Apocynum venetum L.), Nagarmotha (Cyperus rotundus) all show extremely potent anti-glycation activity (Gugliucci et al. 2009; Hsieh et al. 2005;, Ardestani and Yazdanparast 2007 and Yokozawa and Nakagawa 2004).

A ten-fold dilution of green tea, mint, sage, lemon balm and black tea was found to almost completely stop the formation of AGE's. When AGE levels increased to 20 fold, only lemon balm showed the ability to retain its capacity to stop AGE accumulation. The effective anti-glycation capacity of the above herbs is as follows = lemon balm (89.8%), mint (47.8%), black tea (38.0%), green tea (35%), sage (33 %), common verbena (30%), rosemary (18.8) and lemongrass (3.0%) (Ho et al. 2010) (604). An extract made from Snutellaria baicalensis Georgi flowers has been shown to significantly reduce advanced glycation end products (AGE's) (605).

Tips to reduce AGE's during cooking

The Maillard reaction starts at low pH numbers (acidic), then gradually grows, peaking at around pH 10 (O'Brien and Morrissey 1989). If the moisture level is high, natural decreases in the Maillard reaction will occur. High temperature dry heat cooking promotes AGE's. However heating food in an oven that has high humidity with short cooking times, a lower temperature, or by adding acidic substances such as vinegar or lemon juice, reduces AGE's considerably (Finot and Magnenat 1981) (606). The Maillard reaction rate doubles as temperature grows by 10 °C. If browning is used to measure a Maillard reaction, then one month at 20 °C, 3 hours at 100 °C, and 5 minutes at 150 °C create similar results (Ledl and Schleicher 1990). In summary, foods that are dry or low in moisture and cooked at high temperatures create significantly higher AGE's (Uribarri et al. 2010).

Rifampicin is an antibiotic prescription medicine, that exists naturally in the soil of Easter Island. This has been shown to reduce AGE's. Rifampicin has also been found to enhance the lifespan of worms (Golegaonkar et al., 2015) (607).

Chapter 17
Healthy Air for a Healthy Body

The air we breathe contains numerous particles within it, both good and bad. If you have ever walked through a rainforest after a heavy rainfall you can smell all the nutrients in the air as they are released from the soil. If a natural environment contains soil, bark or plants that contains an abundance of substances that have low molecular weights, these particles will rapidly penetrate the blood brain barrier. This has been proven with the substance pinene, which has a low molecular weight. Experiments on mice have shown [608] that after breathing in pinene, that pinene was detected in their brains for up to 30 minutes after and the same research study concluded that levels of α-pinene were 2-fold higher when the pinene was delivered in an oil blend. This was done by allowing pinene to circulate in the open air when it was mixed with limonene, cymene or cineole.

It is my hypothesis that because tree barks contain some of the most powerful anti-aging substances (willow bark is one example), that the reason the substances in certain species of tree bark are so healthy is because they collect and store these low molecular weight anti-aging substances. For example willow bark, from which common household aspirin is made, contains the substance salicylic acid. This means with the right species of tree, you could grow anti-aging herbs and plants in the environment and harvest the bark for health and healing. Salicylic acid is the main source of commercial aspirin and comes from white willow bark and wintergreen leaves.

When crickets were given metaformin, which occurs as Berberine in the shrub Arcangelisia flava (L.) Meer, Oregon Grape (*B. aquifolium*), Goldenseal (*Hydrastis Canadensis*), Chinese goldthread (*Coptis chinensis Franch*) and Indian barberry (*B. aristata D.C*) [609] they lived for between 188 and 194 days, compared to the control group which only lived between 136 and 163 days. When aspirin and metformin were both combined together, the groups lived 163 days. The best results were obtained when aspirin was given in low doses allowing the crickets to **survive at a rate of 234%** longer than the control group (610).

The Willow Bark Formula

I have used the following formula below with success by taking it before bed. The elderberry extract contains cyanogenic glycosides, in which the sugar derivatives can form natural plant based cyanide (as do flax seeds). So use caution, as the elderberry may be overwhelming if used in excess. It is a good formula to help you wake up refreshed the following morning. The formula is as follows:

3 drops of willow bark extract (is used to make aspirin)
2 drops of lovage extract
1 drop of elderberry extract
3 drops of schizandra berry extract
2 drops of cinnamon extract
2 drops of sulforphane extract
1 1/2 teaspoons of brewer's yeast powder

Take the above formula with flax seeds that have been soaking in water for 20 minutes or longer and add a little bit of honey.

Willow Bark and Chlorogenic Acid

The major substances in willow bark (Genus Salix L) and the leaves of the willow tree are salicin, chlorogenic acid, rutin and epicatechin. Trans-cinnamic and coumaric acids also exist, but their levels vary between trees. The plant Creeping cucumber (S. amplexicaulis) contains the highest levels of chlorogenic and coumaric acids and the Purple pitcher plant (S. purpurea) has an abundance of coumaric and epicatechin acids (611).

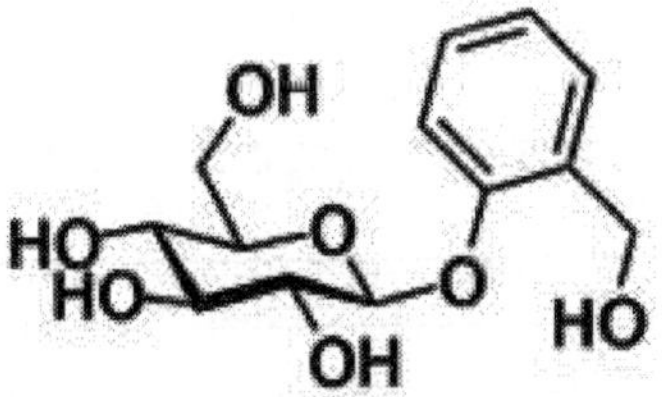

Salicin

The species of willow bark that contains the richest source of salicin is Purple pitcher plant as well as the bark of the Creeping Cucumber (612).

Further reading

Coffee enriched with willow (Salix purpurea and Salix myrsinifolia) bark preparation – Interactions of antioxidative phytochemicals. Agata Durak et al. Oct 2015.

Removing Toxins from the Tissues

When excess heavy metals build up in the body, they act as fuel for the free radical process, which can contribute to cerebral impairments. For example, being exposed to high aluminum levels in drinking water can negatively affect short term memory and a loss of mental concentration (613).

Researchers have found (614) (615) that during times of reduced solar activity, the negative health effects from breathing in excessive air pollution on the body are not as extreme. This effect occurs because pollutants in the atmosphere absorb excessive solar radiation that takes place during higher solar activity, which causes toxic photochemical reactions to take place in the air. This than produces substances that negatively impact health (614) (615).

The best substances for removing global air pollution from the body are - Lysine, Olive Leaf Extract and Salacia reticulta Extract.

A John Hopkins research study (616) (617) found that a broccoli sprout-infused beverage (*abundant in sulforaphane*) reduced the effects of benzene which is one of the most common airborne pollutants. The clinical trial involved 300 Chinese women and men living in one of China's most polluted regions (Jiangsu Province). Sulforaphane has also been shown to protect against diesel exhaust (618).

Another clinical trial (619) examining a combination of mixing brown mustard powder with cooked broccoli found that the sulforaphane in the broccoli was absorbed four times more than eating the cooked broccoli alone. Other substances proven to fight the effects of traffic-related air pollution on the body –

- Vitamin C, fish oil and eating an antioxidant-rich Mediterranean diet (620)
- Schisandra chinensis protects the skin from air pollution (620)
- Cocoa has been shown to help the bodies of young children detox air pollution from their systems (622)

While studies (623) have found that Vitamin E reduces inflammation caused by air pollution, taking higher levels of Vitamin E for long periods of time may actually aggravate the effects of air pollution on the body. For example, in a meta-analysis study reviewing 19 clinical trials, researchers found that high doses of vitamin E (≥400 IU/d) for a year or more may increase all-cause mortality (623).

Black cumin seed with its abundant levels of Thymoquinone, has been shown to prevent damage caused by diesel exhaust particles (624). A research study titled: Citrus Essential Oils in Aromatherapy (625), stated carveol, perillyl alcohol, limonene, carvone and geraniol have been shown to help alleviate carcinogenesis, which is a biological reaction that occurs from breathing air pollution. Carveol is found in abundance in black tea, dill, caraway, mandarin (Citrus reticulata) and blackcurrant berries (Crowell et al., 1992) and caraway based plants have been used to treat various diseases (Mahboubi, 2019). In Traditional Chinese medicine (TCM) Carveol is used as an astringent, antispasmodic and carminative for treating dyspepsia and indigestion (Sachan et al., 2016). It also exerts anti-hyperlipidemic, antioxidant and anti-inflammatory effects in the liver (Johri, 2011). Carveol also exerts

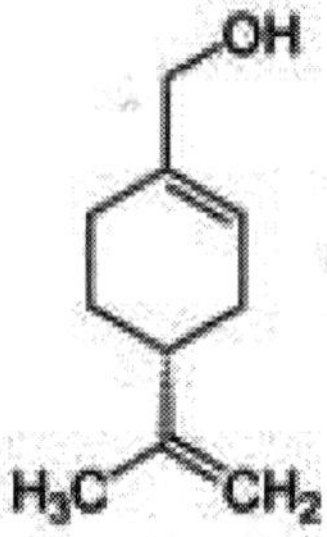

Perillyl alcohol

anti-inflammatory effects comparable to aspirin (Kawata et al., 2008) and reverses motor and sensory coordination neurological problems if taken daily in a dose-dependent manner (626).

Coriander (*also called Cilantro*)

Besides being a room temperature probiotic, extracts of coriander made from its leaves have been used to reduce the levels of lead, mercury and copper in fish and shellfish (627). In regards to Coriander removing lead from the body, a randomized clinical trial that involved 32 children aged 3 to 7 who took Coriander extract for 2 weeks showed a reduction of lead in their bodies (628) and in another study (629) mice that were subjected to lead poisoning had a reduction in lead deposits in their body after eating Coriander. Another study (630) was published showing that Cilantro soup enhanced mercury excretion after dental amalgam removal. A list of plant based metal chelators is shown below -

- N-acetyl-cysteine, especially when combined with vitamin C, is used to remove toxic metals from the body (631).

- Salicylic acid has been used in lead detoxification. (Yang et al. 2004) (632).

- Chokeberries (Aronia melanocarpa L.) An extract made from Chokeberries, which is rich in polyphenols, anthocyanins and proanthocyanidins, was found to remove zinc and copper when the pH was at 5.5. The authors in the study (633) stated this was due to the abundance of Cyanidin 3-O-β-galactoside and polyphenols.

C-reactive protein levels (CRP)

When your body is under prolonged stress and is low in specific nutrients, your liver makes up for the lack of nutrients by trying to fight the stress. This can result in chronic inflammation. However, just before this inflammation occurs, there is an increase in the C-reactive protein levels in your blood. C-reactive protein levels are elevated during above normal solar activity. A 2017 research study (634) looked at the effects of above average solar activity on elderly people living in Boston, Massachusetts. It involved 742 people and the study took place between May 2000 and December 2017. The average age of people in the study was 73 years. The authors of the study discovered that changes in sunspot and geomagnetic activity caused the participant's T-cells, which regulate the immune system, to become over-active. (T-cells help fight off and viruses and help fight disease). The researchers went on to state in their study that above average solar weather conditions were affecting their body's C-reactive protein levels (CRP). C-reactive proteins are proteins in the liver which grow as inflammation begins rises in the body. The authors discovered significant increases in CRP levels depending upon sunspot activity. CRP levels were not affected however by increased geomagnetic activity on earth, only the sunspot activity. The study also discovered the effects were similar to effects on the body experienced during air pollution. What was interesting is earth's geomagnetic activity affected the participant's fibrinogen levels, which was not affected by sunspot activity. Fibrinogen regulates blood clots, with healthy levels showing a person is less at risk for cardiovascular disease.

Another study (635) found that for each increase in CRP levels, there was a 1.4 year reduction in lifespan in people aged 57 years or more. This study shows confirms CRP levels can be used to predict long-term survival in middle aged men.

Foods and Supplements that lower CRP Levels

Olive Oil (636), Potassium (best) (637), Folate (638), Green tea (best) (639), Vitamin E and Ginger (640).

Chapter 18
Exploring Biostimulants In-depth (*plant growth regulators*)

Biostimulants, also called bio-enhancers, are a purified extract / substance that is sprayed onto crops to enhance their growth and crop yield. Bioenhancers include: caraway, polyethylene glycol (peg), salicylic acid, glycine betaine, seaweed extract, chitosan, fulvic acid, salicylic acid, jasmonic acid and safranal or saffron extract. Biostimulants have similar effects to that produced by the Maillard reaction, in which they enhance the absorption of nutrients, or allow for a deeper penetration of such. One of the most widely used bioenhancers is known as Capmul, which is produced from oils and is used in lip products. When the antibiotic ceftriaxone was combined with capmul, it enhanced the gastrointestinal absorption bioavailability of the prescription drug Ceftriaxone by up to 79% (641). Biostimulants also help plants fight drought stress (642). When the amino acid Glycine is applied to crops, it increases their polyphenol levels and antioxidant levels (643). And when the amino acids tyrosine, lysine, and methionine are applied to crops of tomato plants, it enhances plant growth, with tyrosine showing the best results (644).

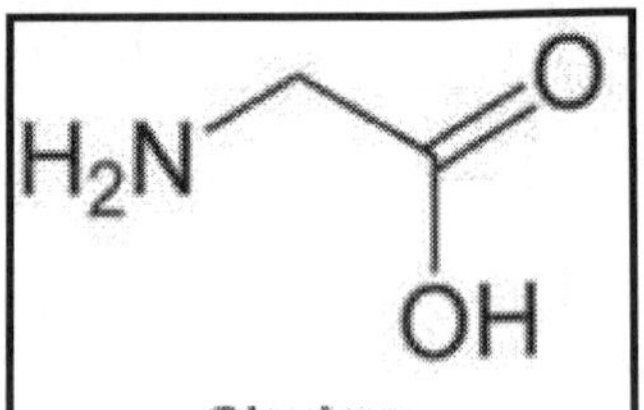

Glycine

Maillard reactions in Wood Chips

Throughout the Alchemical literature there are terms that refer to heat that is long and slow, with some references being made to alchemical reactions occurring in situations where debris is slowing rotting. Another bioenhancer is wood chips. In a research study (645) that lasted from 1951 to 1965 at a soil research farm in Marcellus, NY, the researchers tested a 5-year rotation of tomatoes, sweet corn, beans, peas and cabbage. They tested 14 different types of soil treatments, which included applying 10 tons per acre of wood chips added each year. Other treatments involved testing grasseed cover crops or over wintered ryegrass. The crops that used wood chips showed the best results when the chips had been top dressed on the soil after planting the crops, instead of being plowed under. The study found that nitrogen levels increased as well as soil organic matter in the plots that contained the wood chips. However the plots without the wood chips showed drops in nitrogen and soil matter over the 15 year test period.

This study is quite a breakthrough because it shows that the yearly addition of wood chips maintained better soil quality than by simply using grass cover on the crops. This method was also tested by farmers in Quebec who used hardwood chips that measured 3 inches in diameter, also known as "ramial". Ramial is chipped wood that is made from small to medium-sized tree branches. The farmers in this test also obtained good results, probably because the nutrients in the decaying bark are higher if the tree branch diameter is less than 3 inches. In conclusion, the best way to enhance crop yield would be to topdress the soil at 5 to 10 tons of remial chips each year and then work them into the surface of the soil during cultivation.

Additional Bio-Stimulants

- Salicylic acid at 0.75-mM concentration has been shown to increase root growth in plants (646).

- Jasmonic acid has been shown to improve the yield and growth of wheat in drought conditions (647) as well as improve cotton productivity during drought conditions (648).

Seaweed extract (*when administered at 5%*) has been shown to significantly increase chlorophyll, carotenoids and photosynthetic rates in corm (649) as well as corn crops, which allowed the corn to significantly absorb more zinc, iron, copper, magnesium and silicon as well as many other important nutrients (650). As a matter of fact seaweed extract has been shown to improve just about every crop there is (651) and when Safranal or Saffron extract is sprayed on tomato plants, it results in increased plant height, increased antioxidants and significant increases in the tomato plant's anthocyanins, lycopene, carotenoids, flavanols, polyphenols, flavonoids and condensed tannins (652).

So what do these all have in common? The majority of these biostimulants are all anti-aging substances. Hence these substances, when taken with certain foods, may enhance their nutrient value due to their ability to better absorb nutrients into the body. Biostimulants are well known for their ability to significantly increase the concentration of polyphenols in plants (653). Why is this so? I hypothesize that because most biostimulants are hydrophobic in nature, they don't allow dirt or dust to cling to them, much like oil resists water. This allows the plant to remain free of toxins or dirt that would hinder its maximum nutrient bioavailability potential, allowing for an overall better absorption of nutrients.

Diving Deeper into the Longevity Secrets of Gotu Kola (Centella asiatica)

Gotu Kola is an amazing herb with numerous health benefits. Let's take a look at just a few of the many health benefits of Gotu Kola. Gotu Kola from South Africa was found to contain α-pinene as one of its main constituents (654). It also contained minor amounts of carene and linalool (655). One of the most abundant substances in Gotu Kola is Caryophyllene (656) (657), containing up to 36.3% (658). Caryophyllene has shown significant benefits in reversing age-associated memory impairments in mice. What's very interesting about Caryophyllene is that the memory enhancing effects work better on older mice,

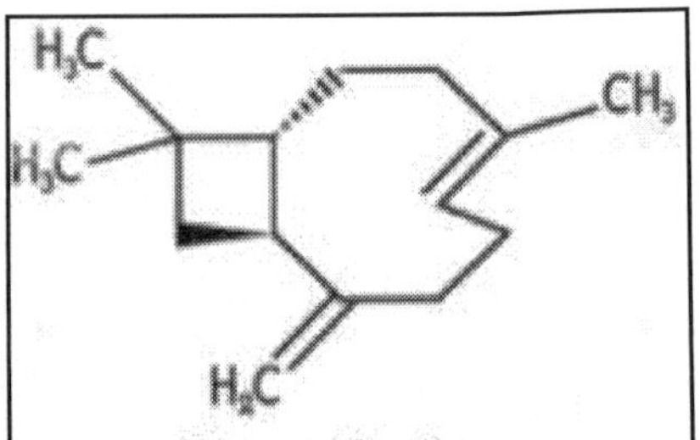

Caryophyllene

compared to young mice (659). Caryophyllene has also been found to extend the lifespan in worms up to 22% (660) as well as exhibit defensive mechanisms that repel insects (661).

Caryophyllene and the tree of Longevity

A tree that contains an abundance of Caryophyllene is the Chinese white pine or Armand pine (P. armandi), which grows at altitudes of up to 3000m (662) (663). The Armand pine tree enjoys growing in sub-alpine and temperate regions of earth's northern hemisphere and lives up to 1000 years (664). The tree is considered an emblem of immortality and longevity by the Chinese. Legend has it that Qiu Sheng who was founder of the Shang Dynasty stated that pine-resin was the reason for his longevity. It is interesting to note that the Shouxing Chinese god of longevity is depicted as standing at the foot of a pine tree with a fairy-crane perching on its branch (662) (663).

Armand pine

Chapter 19
Published studies on substances that treat Glaucoma and Cataracts

The following substances listed in this chapter come from published studies and have successfully been used to treat cataracts and glaucoma. Many of the substances used are dose dependant and also use specific parts of the plant, in most cases the leaves. If you decide to use any of these, I encourage you to look up the full referenced study for complete details on dosage, part of plant used and any other studies or feedback regarding the use of the plant for treating cataracts or glaucoma.

Hachimijiogan [665] [666] is a formula consisting of the following eight herbs -

- 6.0 g of the root of Chinese Foxglove (Rehmannia glutinosa Libosch)
- 3.0 g of Cornus fruit (Cornus officinalis Siebold)
- 3.0 g of the rhizome of Chinese potato (Dioscorea batatas Decne)
- 3.0 g of the rhizome of Asian water plantain (Alisma orientale Juzepczuk)
- 3.0 g of the sclerotium Poria cocos wolf common name (Poria cocos Wolf)
- 2.5 g of the root bark of Tree peony (Paeonia suffruticosa Andrews)
- 1.0g of the bark of Chinese Cinnamon

(Cinnamomum cassia Blume)

- and 0.5 g of the root of Carmichael's monkshood (Aconitum carmichaelii Debeaux).

Goshajinkigan. A diabetic woman aged 65 with a left eye cataract who had tried many western medical treatments, with no success, was given the herbal formula Goshajinkigan in combination with eye drops. The herbal treatment significantly improved her condition. Ten days into the treatment, her visual acuity did not further deteriorate. The authors in the study stated Goshajinkigan may be of use for treating diabetic women diagnosed with cataracts (667).

Quinax (dihydroazapentacen or azapentacene polysulfonate). The authors in this study evaluated the effectiveness of Quinax. The study involved four groups of patients diagnosed with cataracts. The study found that after 5 years of use, it prevented development of early senile cataract and it also slowed down the progress of the disease. The herbal formula was ineffective with patients diagnosed with advanced diabetic cataracts.

Baicalein, ginsenoside, resveratrol, forskolin and hesperidin have all been found lower intraocular pressure (669).

In Ayurvedic medicine, Triphaladi Ghana Vati and Elaneer Kuzhambu Anjana in Timira, exhibit anti-cataract effects. A clinical trial (671) involving patients diagnosed with Senile Immature Cataract who took 500mg of Ghana Vati orally for 3 months, with another group taking Triphaladi Ghana Vati 500 mg orally and Elaneer Kuzhambu Anjana, also for 3 months, exhibited statistically significant improvements in their blurred vision, glare stranth, bright light tolerance, distant visual activity, pinhole vision and corrected visual acuity. The Triphaladi Ghana Vati group who had been diagnosed with nuclear cataract exhibited significant improvements in their night vision. The authors in the study concluded that the formula may arrest, prevent or delay senile cataract.

Saffron Supplement. In a clinical study (671) patients diagnosed with primary open angle glaucoma stated that after taking 30 mg of a saffron supplement daily for four weeks that they experienced significantly reduced IOP levels (Intraocular eye pressure) after three weeks.

A research study (672) screening herbs most used to treat eye disorders were surveyed. Here are the herbs used most often, with the most popular herbs listed first - 18.4% Green Tea (Camellia Sinensis), 10.3% German chamomile (Matricaria chamomilla) and 5.5% Spearmint (Mentha spicata). While green tea was the most often used herb, published studies now confirm that green tea shows significant anticataract potential as well as for being used for the prevention of glaucoma (673) (674) (675).

Green Tea for Good Eye Health

Green tea combined with Ginkgo biloba extract, grape seed extract, alpha lipoic acid, citicoline, Co-enzyme Q10, N-acetyl-cysteine and curcumin was used in a Phase 1 clinical trial (676) in patients diagnosed with open-angle glaucoma diabetic retinopathy. The clinical trial was terminated after the primary researcher quit his job. In another clinical trial (677) that was also terminated mid-way through its testing, (***clinicaltrials.gov identifier: NCT03866005***) green tea was combined with vitamins B1, B12, C, D, E, coenzyme Q10, lipoic acid, lutein, grape seed extract, curcumin, zeaxanthin, resveratrol and Pycnogenol (from the bark of Pine Trees). The participants orally ingested two or four capsules daily and had been diagnosed with diabetic macular edema. Additionally, another clinical trial (677) involving patients diagnosed with diabetic retinopathy, that as of 2021 was still undergoing testing by Western University of Health Sciences Diopsys, Inc. mixed green tea leaves with vitamins C, D3, E, zinc oxide, coenzyme Q10, tocotrienols/tocopherols, benfotiamine, eicosapentaenoic acid, zeaxanthin, lutein, docosahexaenoic acid, α-lipoic acid, resveratrol, turmeric root extract, N-acetyl cysteine, grape seed extract and Pycnogenol. The particpants in the study took two softgels of this combination once a day for a total of six months. And finally another clinical trial (677) that was last updated in 2009 and conducted in Massachusetts, mixed green tea extract (200 mg per day) with Lutein to test its effectiveness on patients diagnosed with age-related macular degeneration (AMD). It is interesting to note here that many of these clinical trail studies never reached their completion.

Green tea contains an abundance of Epigallocatechin-3-gallate (EGCG). A research study found that EGCG significantly alleviated optic nerve injury and stated EGCG was able to attenuate the symptoms of glaucoma (678).

Spearmint exhibits anticataract activity. In a study (679) involving goats who were given cataracts using the glucose method, the study found that an extract from the leaves of spearmint (M. spicata) prevented glucose-induced damage in the lenses of the goats and suggested that the spearmint extract may delay the onset of cataract.

Eyebright. An eye drop solution containing Eyebright herb (Euphrasia officinalis) mixed with Chamomile was found to protect the eyes (*on human corneal epithelial cells*) against excessive UVB rays (680).

Psyllium (Plantago ovate)**.** In a Persian medicine text titled: The Canon of Medicine, that was authored by philosopher Avicenna (980-1038 AD0, it states that Psyllium (Plantago ovate) be used for the treatment of dry eyes. A randomized, placebo-controlled double-blind clinical trial (681) involving sixty patients diagnosed with dry eyes were given Psyllium ophthalmic drops four times daily for six weeks. The study found that Psyllium was an effective, natural and inexpensive method that has beneficial ocular effects on patients suffering from dry eyes.

Sandhu and coworkers (2011) screened 262 plants for various ocular diseases, including different types of conjunctivitis, and discovered 51 were found to help reduce swelling and redness of the eyes (682). The most often used part of the plant was the leaves. The research study lists all 51 plants along with cited references (682).

Extracts made from the leaves of Permotrema perlatum, Merremia emarginata, Tridax procumbens and Euphorbia prostrate were found useful to help prevent the eye from damaged caused by diabetes. (683).

The research paper titled: **Traditional Eye Medicines in Tanzania: Products, Health Risk Awareness and Safety Evaluation** (684) is highly recommended reading as it gives numerous natural plants and herbs, what parts to use, dosages etc. You could write an entire book on all the data in this well published paper. The paper states that the leaves of the plants are the most often used part for treating cataract, and that inflammation of the eye was cited the most often.

Bidens pilosa L

Before you decide to try these remedies, I strongly advise you to do your own independent research to find out further information on how each plant has been used to treat cataracts. You may also want to begin with a more diluted concentration of the substance than increase concentration until you feel you have the best results. Regarding the aforementioned paper, regarding cataracts, it shows how the following plants are used and which parts of the plant are used:

- Microglossa pyrifolia Lam. Root juice is used as

eye drops.

- Dolichos trilobus L The leaves are boiled than cooled and applied.

- Bidens pilosa L. The leaves are squeezed into juice and applied at 3 drops three times daily.

- Vernonia amygdalina Del. The application is the same as Bidens pilosa.

- Commiphora edulis. The fresh latex is used after leaves have detached from the stem. It is applied daily until recovery.

- Phyllanthus amarus Schum. & Thonn. The same same application as Bidens pilosa L

- Euphorbia hirta L. The fresh latex is used after leaves have been detached from the stem. It is applied daily at 3 drops daily.

- Swietenia macrophylla King. The fresh latex is used from its bark with one drop applied once a day.

- Rumex usambarensis (Dammer). Aerial parts of plant are squeezed, than juice is used as an eye drop applied twice daily until recovery.

- Solanum lycopersicum L Its fresh leaves are squeezed and the juice is applied as an eyes drop.

Ceasalpinia digyna and Alangium lamarckii. When used in alcoholic extracts, this plant was found to significantly reduce aldose reductase (AR) levels [685]. Aldose reductase is a condition that contributes to cataracts in patients that have been diagnosed with diabetic eye diseases [686].

Allium cepa L. (Tuber/bulb). Onion juice at 50% dilution was found to prevent selenite-induced cataract formation by 75% in mice (687).

Vitamin C. Studies by Jacques et al., (687) found people that had high levels of vitamin E, vitamin C and carotenoids in their bodies exhibited a reduced risk of cataract. Vitamin C has been shown to prevent selenite-induced cataract formation in mice up to 41% after using it for two weeks (687).

A research study (688) found the following prevented cataract formation, showing the most effective substances first - **Propolis 70%, diclofenac 60%, vitamin C 58.4% and quercetin** (found in abundance in onions) **40%**. The study also concluded that Ginkgo biloba (Egb 761) extract had no effect on helping prevent the formation of cataracts. Because diclofenac is a prescription pharmaceutical, the Chinese formula Duhuo Jisheng Wan has shown similar results to diclofenac, although no studies have yet been conducted on using it for cataracts (Chinese herbal recipe versus diclofenac. Supanimit Teekachunhatean et al. Dec 2004).

Caffeic acid phenethyl ester. Caffeic acid phenethyl ester was found effective in suppressing cataract formation in the eyes of mice (689). Caffeic acid phenylethyl ester was also used to reduce intraocular pressure in mice by exhibiting significant protection of their ganglion cells, which are cells in the eye (690). Caffeic acid phenethyl ester is found in abundance in propolis (691) and propolis is used to help remove wrinkles (*Bee Products in Dermatology and Anna Kurek-Górecka, et al. Jan 2020, Comparative evaluation of propolis. Ola M. Elkhateeb et al. Oct 2022.*).

Silk Cotton Tree (Cochlospermum religiosum). These leaves significantly reduced selenite cataract via its strong antioxidant ability taking place in the rat lenses (692).

Little Ironweed (Cyanthillium cinereum). This plant has shown protection against atherosclerosis and diabetes (693). When lupeol was isolated from its leaves, it was shown to protect against the formation of nuclear opacity in mice (694).

Lupeol which was isolated from Vernonia cinerea, was found to stop the formation of cataracts in mice (695).

Oregano (Origanum vulgare L). This spice exhibited significant protection against mice diagnosed with selenite induced cataract (696).

Indian gooseberry (Phyllanthus emblica). In a study on mice (697) who received 2% Indian gooseberry and a 0.2% standardized mixture of its tannoids, the researchers found that it was effective in retarding the development of their diabetic cataracts.

Byakangelicin. This substance which was isolated from the roots of Angelica dahurica, and administered to mice (at a dose of 100 mg/kg for two weeks) who were fed a diet of 30% galactose in order to induce cataract formation, experienced a significant reduction in cataract formation (698). **Byakangelicin** is also used to treat headache, colds and toothache (699) and is also found in Limeberry (Triphasia trifolia), Curry Leaf Tree (Murraya koenigii) (700) and Lemon Marmalade (701). While no studies have yet confirmed Byakangelicin's ability to upregulate the SOX9 gene, it has been shown to exert anti-osteoperatic effects and prevent bone loss in studies on mice. As shown throughout this book, many substances that stimulate SOX9 also have strong cartilage rebuilding and osteoarthritis prevention properties (702).

Chinese chaste tree (Vitex negundo L). Flavonoids from this plant species were found to protect against selenite-induced cataract via its antioxidants (703).

African Tulip tree (Spathodea campanulata P.Beauv). The brilliant red flower buds from this plant are used in Nigeria as a local eye drop to improve vision. Drops from the African tulip tree exhibit successful anti-cataract activity effects similar to the FDA approved prescription drug Captopril (704).

Rambutan (Nephelium lappaceum). A South Korean patent (#*KR20210142068A*) states that an extract made from the seeds of this tropical tree are used to treat cataracts and age related diseases (705).

Japanese Cranesbill (Geranium thunbergii). A substance known as geraniin, which is a tannin that is isolated from this plant, was shown to exhibit protective effects in the eyes of mice (706) as well as prevent advanced glycation end-products (AGE) by 43% (Rambutan rind in the management of hyperglycemia. Uma Palanisamy et al. Jan 2011).

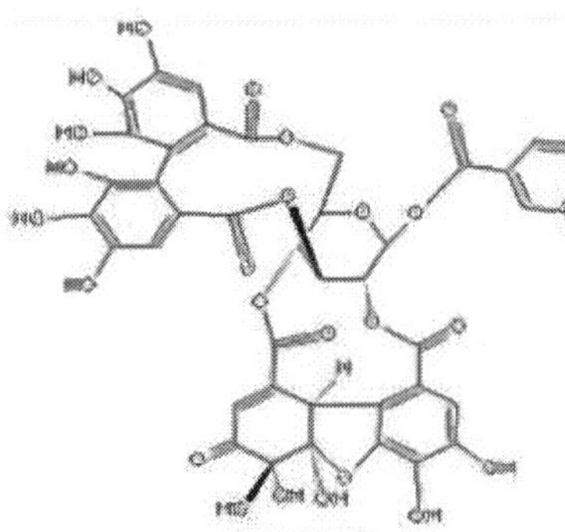

Geraniin

Silk-cotton tree (Cochlospermum religiosum). A 95% ethanolic from its leaves exhibit anticataract activity (707).

Rosmarinic acid - found in Chinese clary (Salvia grandifolia), Taro (Colocasia esculenta), the leaves of Beefsteak plant (Perilla frutescens) and Self-heal (Prunella vulgaris). Exhibits significant effects (707) all show good anti-cataract activity.

Self Heal

Isorhamnetin3-glucoside (found in the leaves of Cochlospermum religiosum and onions) was found to significantly retard selenite cataract in vitro (708).

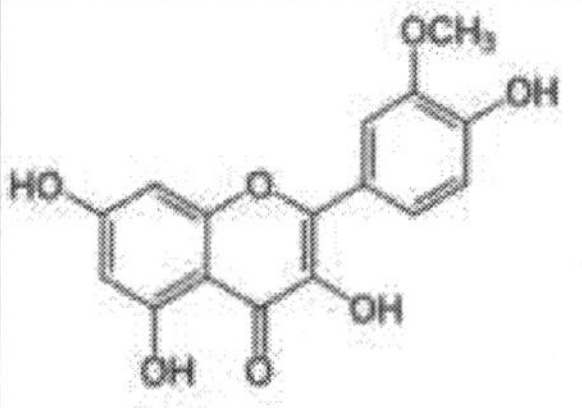

Isorhamnetin

Diosmetin, which is the main component in the Chrysanthemum flower, was found to protect the retinas of the eyes against injury (709) and Diosmetin has been used to help treat cataracts (710). Chrysanthemum flower also contains isorhamnetin 3-O-b-D-glucoside, quercetin, apigenin and luteolin (711). Diosmetin is found in abundance in lemon and bergamot juices, which are also abundant in apigenin (712).

Rutin (quercetin-3-O-rutinoside) is a citrus flavonoid glycoside found in abundance in buckwheat (713). Germinated extracts of buckwheat lower blood pressure (713) and studies have found it reduces spatial memory impairment (714).

In a recent March 2021 research study titled: Maillard Reaction Products in Gluten-Free Bread Made from Raw and Roasted Buckwheat Flour, that was published by Maria Barbara Różańska and colleagues, the authors looked at the effects the Maillard reaction had on raw common buckwheat and roasted buckwheat. **The roasted buckwheat exhibited significant increases in catechins and gallic acids, as well as 42 to 46 times higher levels of quercetin in its crumb and crusts**. The crusts and crumbs of the roasted buckwheat also showed a 158% and 187% increase in phenolics (156 and 140 μg/g of crumb and crust), compared to raw buckwheat flour. However the opposite antioxidant effect was observed in Tartary buckwheat. Buckwheat is a much understudied substance with many amazing health benefits. Expect to see more studies on this in the future.

Rutin is also found in blackberries, raspberries and bottled black tea, with significant amounts found in black olives and capers (715). Rutin also increases Akkermansia levels, which are the good bacteria in your stomach (716). A research study (717) administered mice extracts of Ginkgo biloba, rutin and quercetin to the lens' of mice with cataracts. The study found that all three of these substances exhibited excellent results by enhancing glutathione levels, with quercetin exhibiting the best results. The researchers concluded that these results were from quercetin's abitlty to fight AGE's (717). Bergamot is abundant in rutin (718) and bergamot has been suggested for the treatment of cataracts and glaucoma (719).

Japanese Thistle (Cirsium japonicum). Flavonoids extracted from this plant, such as quercetin, have been shown to help inhibit aldose reductase activity, especially when the flavonoids were harvested from the plant during the season of spring (720). Aldose reductase plays a role in the progression of cataract and retinopathy eye disorders of patients diagnosed with diabetes. Japanese Thistle also contains diosmetin (720), which has been proven to prevent cataracts (721). The plant also contains silicristin and silydianin (721). Diometin, silicristin and silydianin all exhibit natural synergy with one another (722) which help the body's cells fight free radicals.

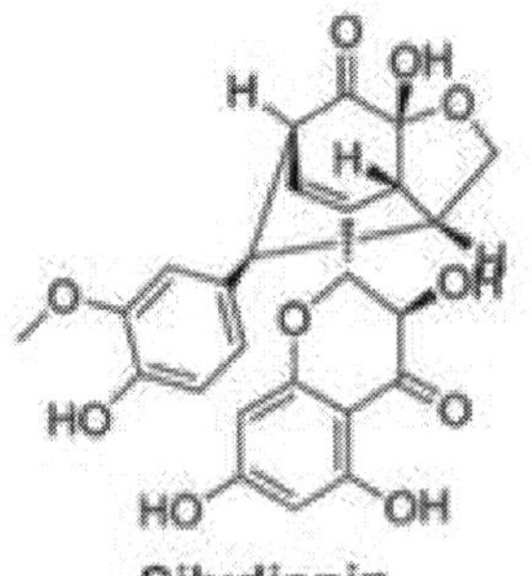

Silydianin

Rutin #2. A combination of rutin and forskolin decreases discomfort in patients with glaucoma suffering from dry eyes (723) and a combination of rutin and aescin has been used to enhance blood circulation in the eyes (724). Other studies state rutin exhibits protective effects against glaucoma in mice showing statistically significant effects (725).

Buddleja Flower (Buddlejae Flos). Out of all the substances tested for Aldose reductase activity (726), a 70% methanolic extract from the Buddleja Flower showed the best results. Buddleja Flower contains ample amounts of apigenin, rhamnopyranosyl, luteolin and luteolin.

Chlorogenic acid has exhibited therapeutic effects for preventing cataracts in diabetic mice (727).

AGE's and Cataracts. A study (728) that looked at the best substances that inhibit cataracts formed via AGE's stated the ones that showed the most protection were: Ellagic acid (exhibited stronger effects than quercetin), Kaempferol, Palbinone, Desmethylanhydroicaritin, Geraniin, Luteolin, Sulfuretin and Glucogallin. The study (728) also lists the plants and herbs that these substances can be found in.

A study listed the following substances as being useful in helping to treat or prevent cataracts. An eye drop consisting of **salted lemon juice**, **Astaxanthin** (0–1 mM) or methanol-water extracts of **Citrus Peels**, (100 and 200 mg/kg body weight) or **Ginkgo biloba** leaves (*protection from radiation induced cataracts*) (729).

Kumquat Peel. Kumquat peels heated in hot water (F. japonica) were found to contain an abundance of rutin with the highest levels found in dried kumquats, with flavonoid levels being higher in unripe kumquats. Kumquat peels contain diosmin, rutin, naringin, neohesperidin, hesperidin, quercetin and kaempferol, with the levels varying between species and region where grown. Kumquat of the Calamondin species in an 80% methanol extract contained hesperidin diosmin and isorhoifolin (730). Kumquat varieties that show an abundance of diosmin are Meiwa Sweet Kumquat (Fortunella crassifolia) and Bitter Orange (731).

Honey from the Stingless bee *(Meliponini)*. These bees are prevalent where I live in Queensland. Their honey has been used to treat cataracts in traditional medicine (732). Studies on mice who received four derivatives of luteolin from honey made by the sting-less bee experienced a significant inhibition of cataract formation

(732).

Ayurvedic Eye Paste. A special natural eye-paste was applied and exhibited statistically significant effects in preventing glucose-induced cataracts (733).

A published study (734) recommended the following combination in people aged 40 and older to protect and strengthen their vision -

Curcumin	800 mg	Coenzyme Q10	100 mg
BPF	250 mg	Zinc	60 mg
Vitamin C	300 mg	PUFAs	1-2 mg
Quercetin	15 0 mg	Vitamin A	0.5 mg
Grape seed extract	100 mg	Lutein	10 mg

An oral Ayurvedic Herbal Supplment for Treating Blurred Vision

Patients took the herbal mixture known as Punarnavashtaka Kwatha and the herbal mixture known as Gokshuradi Guggulu for approximately two months. They than experienced significant reductions in their blurred vision (735).

Limonene reduces the progression of cataracts and significantly reduces glycated protein levels in studies carried out on mice (736) (737).

Aminoguanidine combined with limonene, creates a synergy, showing an almost twenty fold reduction in AGE's (738).

Warnings and Cautions

While many of the substances mentioned in this chapter are from published studies, there may be some instances where applications of facial creams may cause eye problems. For example, there exist

2 case studies where woman were injured when they applied Vicks VapoRub to their bodies. In one case it caused thermal ocular burns (739) and in another case, it was applied to the eyelids and forehead (740) for relief of headache which also resulted in significant eye problems for the patient.

The Eyejuice Formula

I have used the following combination for many years and found it of considerable use for maintaining strong and healthy eyes. Combining into a hot tea - 2 sliced dried Apricots (they must be sliced), 1/2 teaspoon of Coconut Oil, 2 teaspoons of Goji berries, ½ teaspoon of Gotu Kola, a pinch of grape seed extract powder, 3 drops resveratrol extract (or a pinch of resveratrol powder form a resveratrol capsule), 1 vitamin C tablet (*adding sesame seeds is optional*). After bringing water to a boil and adding the above, allow to soak in the cup for 5 minutes than strain and drink. You may want to add some honey, allulose or stevia to sweeten. This formula works well because the coconut oil releases the natural carotenoids from the apricot and goji berries. Even better results are obtained when you eat a handful of goji berries after drinking the tea.

Closing Summary

It is my hypothesis that the red powder of the Philosopher's Stone is a bio-stimulant, that when added to red wine, enhances the body's ability to absorb the longevity nutrients in red wine. When the red power becomes heated to just the right temperature by adding it to a ball of wax than throwing it into molten lead, it transitions lead into gold via the Maillard reaction.

Because the red powder was highly purified, it easily attached itself to the liquid lead. We see this in the case of anthocyanins as they easily attach themselves to lead to remove lead from the body, and in a research study that used anthocyanins from the Dendrobium officinale flower, the anthocyanins extended lifespan 56%. It also doubled the activity of superoxide dismutase levels in worms (353). The extended lifespan published in this study is far above average, and polyphenols when undergoing an increase in heat (Maillard reaction) will show increases in their polyphenol contents.

For example a research study (741) used what's known as the "*smoking process*" to heat Shanxi aged vinegar and increase the polyphenol contents. The process involves a smoking furnace that has a fire below it with a smoking pot located on top. Next a mash formed by the alcoholic fermentation of rice hull, millet bran, wheat bran and other items are placed into the pot which than starts smoking. This is transferred to an adjacent pot every few hours to allow the mixture to be evenly heated. This creates a beneficial Maillard reaction and results in enhanced polyphenols. Results from the study found that smoked Shanxi aged vinegar exhibited increases in the polyphenols ferulic acid, vanillin and vanillic acid, with significant increases in vanillic acid and furanone occurring after just 5 days. The overall polyphenols reached a peak after 5 days, than started to decline with the water-soluble polyphenols increasing up to 12.24%. During the process, amino acids showed decreases of 33.63%. The researchers stated the reason the polyphenols increased was because increases in the temperature caused specific proteins to denature, which in turn allowed for more of the amino acid tyrosine to cause an increase in the synthesis of the polyphenols. Also the increased temperature allowed for breaking of the noncovalent and covalent bonds (*such as hydrogen bonds*) between melanoidin and polyphenols, which promoted a release of polyphenols.

This same reaction was also observed by scientists (742) who heated cocoa, which is an extremely rich source of polyphenols. They concluded that the increase in polyphenols was from the reaction of the amino acids alanine and tyrosine. The authors concluded that the best pH for creating a beneficial Maillard reaction in cocoa is between 7.0 and 7.5 (between 43 and 58 g kg). What is interesting is when cocoa is mixed with almonds and dark chocolate it results in significant reductions in small dense low-density lipoproteins (*Effects of Dark Chocolate and Almonds on Cardiovascular Risk Factors. Yujin Lee et al. Nov 2017*). Lipoproteins are primarly responsible for coronary heart disease. What is also interesting, is that if you ever decide to make your own chocolate, that when plain chocolate was in the pH range of between 4.75 and 5.19 that the chocolate tasted bitter and bland. However when the cocoa was made with the pH level between 5.20 and 5.49, the chocolate tasted delicious.

When green quinoa malt was roasted at 145 °C for 30 minutes, it showed increases in polyphenols (743). However when the temperature was increased to 190 °C, it reduced the levels of polyphenols.

Meteorites and Gold Formation

Research in the future may reveal that the substances used to synthesize gold nanoparticles may give a clue to the substances that create the red powder. It would be interesting to test many of the substances discussed throughout this book via the Maillard reaction when exposed to liquid lead.

The amino acid norleucine is found in meteorites (60) and research shows that meteorite bombardment is responsible for the gold on earth as reported in the September 2011 edition of New Scientist Magazine (*Meteorite storm showered planet in gold. Michael Marshall.*). Isoleucine is in the class of amino acids called Branched Chain Amino Acids (BCAA's) and scientific research shows that when the amino acid isoleucine is removed from the diet, it extends lifespan (98). Perhaps by studying how meteorites create gold on earth, we may better understand how the red powder of the Philosopher's Stone is made.

In final summary, perhaps biostimulants, which have been proven to enhance polyphenols levels in crops, especially glycine, (744) are acting as a conduit for celestial and cosmic forces. We know that amino acids react with lead by attaching themselves to the lead. For example, spraying the leaves of crops with a biostimulant reduces the amount of lead in its roots up to 73% (745).

Rudolf Steiner who founded Waldorf Schools, stated cow horns acted as a cosmic antennae that received and focused cosmic forces and concentrated them to the substances that were placed inside of the cow horn (746). And as covered earlier in Chapter 9, gold can be made from the urine of cows. The biodynamic preparations known as 500 and 501 are placed into cow horns and buried for a few months, than retrieved and diluted with water to create a homeopathic solution that is applied to soil or crops to enhance leaf or root growth. Steiner stated the elements in these preparations carried cosmic and terrestrial forces which imparted their energies to the crops. Steiner also stated plants never suffered from diseases, but were the victims of when "moon influences" became too strong in the soil (Smith and Barquin, 2007) (746). He used weak infusions of dried horsetail (Equisetum arvense) to treat many crop and soil fungal diseases (Preparation 508). In the case of pests, Steiner recommended catching the weed, insect or rodent and then burning it. These ashes were then scattered over the soil to prevent future infestation (746), with the best results being while the planet Venus was rising when the Moon was in Scorpio. There are now some good Antipodean Astro Calendars online that can give you more information about this.

Some biodynamic gardening calendars use specific moon phases to plant crops. This is based upon a Moon Void of Course Calendar, which basically gives the time that the moon rises in the east with a specific constellation close behind it. My book titled: Farmer's Almanac Secrets. Correspondences of Feng Shui, Lunar and Alchemy, goes into more detail about this. However in final closing, perhaps when the moon is in a specific sign (or a full moon) that the red powder of the Philosopher's Stone is placed into a ball of wax and added to molten lead, allowing it to undergo the Maillard reaction, turing lead into gold. Perhaps the red powder was a type of purified polyphenol or anthocyanin which caused an increase in ionization energy that allowed the lead to turn into gold. As technology becomes more advanced, we may be able to witness the effects of heat when the moon is in a specific constellation or at a specific location in the sky and its effects on the chemical reactions taking place in plant life.

Glycine

It makes sense to create gold from amino acids using cows. The amino acid that is in the most abundance in cow urine is glycine [267] and vegans contain more glycine in their plasma (Plasma concentrations and intakes of amino acids in male meat-eaters, fish-eaters, vegetarians and vegans. J A Schmidt et al. Sept 2015). It is also possible to leach gold and copper using alkaline glycine (Enhancing the leaching rate of gold in alkaline glycine solutions. E. A. Oraby and J. J. Eksteen. World Gold 2019). Hence, the amino acid Glycine may play an important role in the making of the Philosopher's Stone.

In a research study that was published in January 2003 titled: Occurrence of **platinum, palladium** and **gold** in pine needles and conducted by G Dongarrá and colleagues, found that pine needles from trees growing in the urban area of Palermo in Sicily, Italy contained **Platinum and Palladium** and that Gold was also found in the pine needles **accompanying the the Platinum and Palladium**.

The researchers stated the most probable source of these elements was urban traffic. However, I hypotheise that the gold that is gathering in these pine needles is not coming from the traffic, but from the sunshine or the air. As stated at the beginning of this book, Sabine Baring-Gould stated that the effects of the Philosopher's stone could be traced to life giving properties of the sun [265]. Another fact to support this hypothesis is that eucalyptus leaves contain gold (Discovering gold in gum trees. Melvyn Lintern et al. Oct 2013), although the authors in this study state the gold is being drawn up the roots from the gold in the soil. Further research is necessary to confirm this hypothesis.

Just as gold nanoparticles are starting to be made from various substances, certain substances are easier than others to make the red powder. This effect is similar to the herb gotu kola, which shows substantial increases in its polyphenols when heated compared to other herbs. There is one way to determine if gold has been made by the Alchemical method. The resulting gold will be ultra pure; the highest quality gold. This also means that the red powder has purifying properties; it is purified via a metal chelation type process. For example, anthocyanins will attach themselves to lead to draw it out of the body. As covered in Chapter 19, when anthocyanins from the Dendrobium officinale flower were purified, they extended the lifespan of worms 56% (353).

Further Reading

COVID-19 and Diabetes Mellitus: May Old Anti-diabetic Agents Become the New Philosopher's Stone? Theano Penlioglou et al. June 2020.

Chapter 20
Exploring the Powers of QI Gong for Longevity and a Healthy Life

My research over the years has led me to the conclusion that as a person ages, the first signs which contribute to ill health is a lack of good circulation in the body; especially the blood vessels. For example, scientific studies show that vascular calcification is the result of poor diet and a lack of exercise (747). Vascular calcification is a condition where calcium forms hard crystals in the walls of blood vessels causing the arteries to harden. One simple, effective and low-cost approach to healthy circulation is long distance hiking or walks, especially after a period of heavy rain.

Over the course of thousands of years, the Chinese have mastered the art of Qi Gong. There are now well published studies showing QI Gong breathing methods significantly promote healthy microcirculation throughout the whole body, with the most recent study being published only a few months ago in March 2023 (748), stating QI Gong and associated breathing may normalize circulation in the body, helping reduce inflammation accumulated in the tissue and help restore healthy metabolism to the body's cells and tissues.

Qi Gong practices such as Opening the Energy Gates Qigong, Dragon & Tiger Meet Movements and Cloud Hands Movements are just a few good examples. If one wants to practice the internal arts, than Baguio Zhang, Xingyiquan and Taijiquan are some good practices to create good inner circulation.

Proper breathing allows one to enter the correct meditative state which in turn allows one to protect themselves and their environment from premature aging, illnesses and to ensure wellness. Many of the methods from which QI practices originated came from observing the cycles of nature, especially those animals and plants that live in extremely harsh environments.

All Taoist and Internal Alchemy practices stress that breathing is key. This is because as you breathe, waste gases are removed out of your body through what's known as gas exchange. This is the process where as you breathe in, the new oxygen is exchanged for the waste carbon dioxide, which the bloodstream carries back to the lungs where it becomes removed from your bloodstream, then exhaled. The ancients state that the health-promoting "Qi" comes from air. The Daoists developed specific exercises which extract "Qi" from the air, and then were able to collect and store it inside the body.

Qi Gong Enhances Alpha Brain Waves

When Qi Gong experts increased their emission of QI energy, they were found to emit infrasonic sounds peaking between 8 Hz and 12.5 Hz (749). This is the same frequency as alpha brainwave activity. As shown earlier in this book, between the ages 60 and 90, there is a major decrease in fast alpha brainwave activity (35). Perhaps the reason people who practice QI Gong are healthier is because they are able to generate healthy levels of Alpha brainwaves.

Using Earth's Natural Frequency to Heal

When Seto et al. measured magnetic field levels that existed adjacent to "Qi Gong" practitioners' hands, he observed a 4Hz to 10 Hz magnetic field frequency with peaking magnetic field values of between 2mG and 4 mG (749). Because the human body exudes magnetic fields of less than 10 to 6 gauss, it shows an increase in magnetic field strength that is 1000-fold higher than usual. Earth's geomagnetic field strength is between 0.3 and 0.6 gauss (750), which also exists in earth's atmosphere (750). Thus it would make sense that generating an artificial magnetic field of these values would generate beneficial healing effects for the body. Let's look at the data for clues.

When a 74 year old man diagnosed with Parkinson's disease was exposed to the frequency of 9.6 Hz for 45 minutes at 0.6 gauss (*applied to the back of the head*), he experienced dramatic improvements in his sleep, balance, strength, motor coordination and speaking strength (750). Eventually his tremor became significantly reduced to nearly nonexistent.

Additional observations of varying magnetic field intensity present in QI Gong Masters (749)

- The frequencies of 4Hz and 10 Hz were present near the palms of a QI Gong master during his "Qi" emission. The data showed the QI Gong master emitted strong bio-magnetic fields of between 2 m Gauss and 4 m Gauss.
- Increased infrasonic sound signals above 70 dB between 8 Hz and 12.5 Hz from a Qi Gong master (*same as brainwave alpha waves*).
- Above average levels of magnetic fields emitted during "Qi Gong" breathing practices.
- Far-infrared radiation was seen being emitted from the palm of a Qi Gong master at a frequency of 0.3 Hz.

Also it is interesting to note that the gastric basal rhythm of the human stomach has a frequency of 0.05 Hz (751) (752) which also happens to be the frequency of earth's atmosphere-generated seismic noise (0.001 Hz to 0.05 Hz) (753). We really are connected to the earth in ways we can't even begin to imagine!

A list of Qi Gong Exercises

As covered earlier, healthy circulation leads to a healthy body and mind. The following exercises are some of the very best to gain healthy circulation in the body-

Qi Gong. Bao Jian An Mo Fa. This is a type of self-massage involving movement and tapping exercises from the martial arts tradition and Chinese Folk Medicine

Taiji QI Gong. These exercises are performed individually as an exercise series. They can be done sitting or standing.

Tree QI Gong. Li Shu QI Gong. This exercise is an energetic practice that exchanges QI with trees.

Zhan Zhuang Gong. Standing pillars. This exercise involves pole exercises or standing in order to create health and strength.

Jing Gong / YI Gong. Shen Gong. These are silent meditation exercises using the imagination and visualization.

Tong Dian Guan Di Fa or San Cai Gong. These exercises help strengthen the connection to the cosmos.

Sun Qi Gong / Tai Yang Gong. This is an inner spiritual and healing program taught by Master Li.

Chapter 21
Miscellaneous Anti-aging information

This chapter is filled with interesting bits of valuable information that don't fit anywhere else in this book, with information only recently made available during the last few years. Before we start, let's close with some thoughts on Carene.

Final closing thoughts on Carene

I strongly believe that in the future, Carene will play a major role in anti-aging science. This is because carene is found in above average amounts in the bodies of centenarians (25) and it also exists in large amounts in the longest lived trees that live for thousands of years (20). Carene is also found in lemon, and lemon greatly reduces AGE's (342) and as shown throughout this book, a diet lower in AGE's extends lifespan, especially through caloric restriction (79).

The concentrated form of carene is classified as an irritant. Therefore the best way to get carne is to eat foods, and especially drink puree or juices that contain an abundance of it. This allows your body to metabolize it properly.

Carene is part of the terpentine class of substances which penetrate deeply and terpenoids also act as a form of airborne self-defense by plants, protecting them from insects and other pests (751). If you have ever used a turpentine based wood preserver, you know that it sinks deeply into the wood. This explains that as a person ages, the ability of their body to absorb vital nutrients (*bioavailability*) becomes reduced and that taking foods or substances that have enhanced bioavailability or can carry the nutrients deeper into the body allow for better health. Anthocyanins are exerting similiar effects.

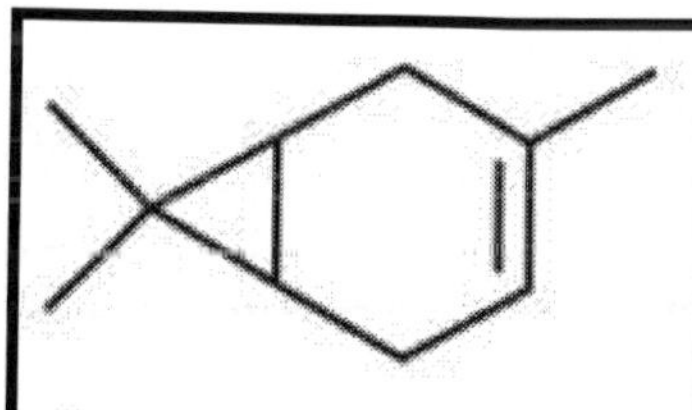

Carene

The Pyramid to Longevity

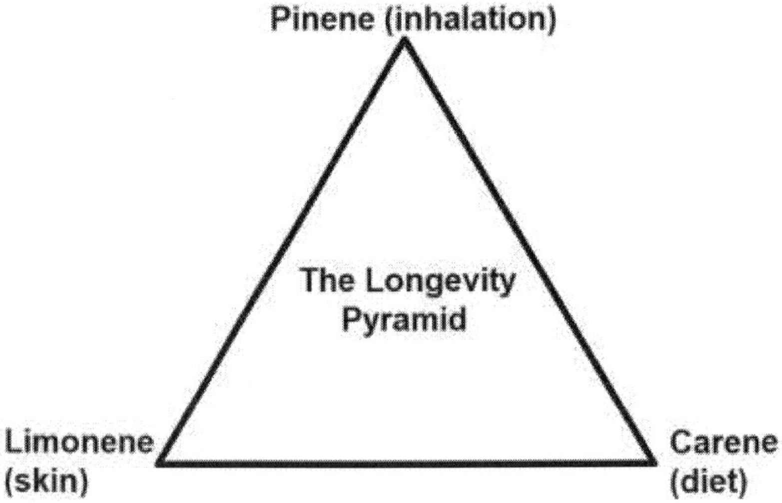

Pinene – Inhalation

Terpenoids work great when allowed to circulate in open air because they are able to deeply penetrate the blood brain barrier. When mice were exposed to oil from the Aleppo pine (Pinus halepensis) for 3 weeks for 15 minutes each day using an electronic Aroma Diffuser (Oregon Scientific WS113), the mice were protected against Aβ1-42 hippocampal damage, which was designed to mimic the effects of Alzheimer's [755]. The researchers concluded that essential oil from the Aleppo pine may play significant roles in neurodegenerative diseases such as Alzheimer's. The authors in the study also found that the main compound responsible for the protective effects was Caryophyllene. Aleppo Pine also contains an abundance of Pinene [756].

Pine needles contain an abundance of antioxidants. When extracts were made from the pine needles of the Lodgepole pine (P. contorta var. murrayana), a tree which lives as long as 400 years, it showed high antioxidant levels, followed by P. nigra var. caramanica [757] and when olive fruit flies (Bactrocera oleae) simply inhaled α-pinene, they showed increased lifespan (758). Pinene is also found in olives and olive leaves (758).

Carene - Diet.

Besides its abundance in pine trees and possibly bark and needles, carene is synthesized by using the metal catalyst **palladium** to reduce its double bond so that it may form a single bond. It is key to note here that purely concentrated synthesized extracts of carene are classified as an irritant. This means it is best taken in foods that contain an abundance of carene or if using it on the skin, used in a diluted carrier such as soap. For example the pine tar bar which contains 1% pine tar (Ego Pharmaceuticals Braeside, Victoria, Australia) is used to treat inflammation associated with numerous skin disorders such as insect bites, dermatitis, eczema, psoriasis, sunburn, nappy rash, chickenpox, urticaria and heat rash (759). Carene as well as Pinene are found in grape plants (Vitis vinifera cv. Malbec), with higher concentrations found in younger plants (*up to 45 days old*) (760). Carene also stimulates NRF2 as well as reduces levels of amyloid-beta in cultures of Alzheimer's disease. Carene as well as α-pinene, β-pinene and limonene have been found to inhibit the bacterial growth of E. coli (762)

Sources of Natural Carene (763) –

- The bark oil of Eldarica pine (P. eldarica), which contains 7.9% caryophyllene and 10.7% carene.

- The leaves of Eldarica pine contain 17.1% caryophyllene, 11.8% pinene, 7.9% pinene and 6.3% carene.

- Turpentine oil from Russian species of Eldarica pine contain 31.5% to 64.2% a-pinene, 2.5% to 11.7% β-pinene and 16.9% to 42.8% carene.

- The bark from the Afghanistan pine (Pinus

eldarica Medw) contains 7.9% caryophyllene,
24.6% pinene and 10.7% carene.

Essential oils that have been extracted from the Maritime Pine tree (P. pinaster) are used in natural herbal remedies as well as a nutritional food source to help assist in the healing of degenerative and chronic diseases (764). Essential oils extracted from the Maritime Pine or its needles are used to flavor food and in beverages, as well as in cosmetics (Maimoona et al., 2011) (764).

While no studies have been done to find out how much carene exists in commercial supplements of concentrated pine bark, extracts such as Oligopin® and Pycnogenol® studies do show that Pycnogenol helps stop beta-amyloid apoptosis of neurons as well as enhances the health of the walls of blood vessels which weaken as a person ages. Pycnogenol has also been studied for its protection against oxidative damage. In a research study where people took Pycnogenol for six weeks, they experienced significantly increased levels of natural antioxidants in their bodies (ORAC), as well as reduced levels of bad cholesterol (765). And another study found that Pycnogenol stimulated the synthesis of antioxidant enzymes inside arterial cells and prolonged the lifetime of natural antioxidants in the body (767). Research studies on the commercial extract called Oligopin® have found that it improved antioxidant levels in women as well as enhanced the ability of their bodies to create

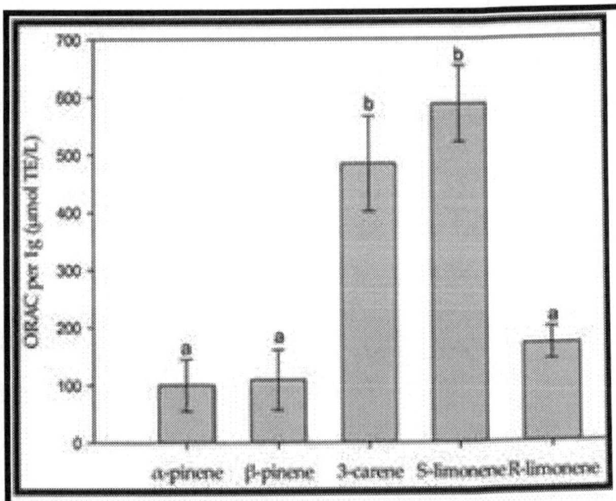

Antibacterial and Oxidative Stress-Protective Effects of Five Monoterpenes from Softwood. Riina Muilu-Mäkelä et al. June 2022.

stronger bones (767) and carene has been found to exert a sleep-enhancing effect when taken orally, which is why pine needle tea made from Afghan Pine (Pinus eldarica) significantly increases the duration of sleep ($p<0.05$) in mice (769).

When a research study (770) tested the antioxidant strengths of pinene, carene and limonene, researchers found that limonene showed the strongest antioxidant activity, followed by carene. The researchers also stated that carene's unusual light absorbance curve is due to its ability to absorb light at 600 nm. This could be why when cells are exposed to 628 nm light, there is an increase in antioxidant genes and mitochondrial respiratory activity, as well as reductions in stress-related genes (771).

Boswellia dalzielii is an aromatic plant whose leaves are used for flavouring beverages (772). Its resin is used as an incense as well as for treating osteoarthritis of the knee (773). When an essential oil was made from Boswellia serrata, it exhibited significant cytotoxic effects on cancer cells. Boswellia dalzielii has been found to contain as much as 27.72% carene and 15.18% α-pinene (774). Boswellia dalzielii also contains the substances rutin, incensole sitosterol and resveratrol (776). This is an amazing plant, so expect to see many more studies regarding the potential health benefits of this plant in the near future. And as a final note, pine oil has been found to significantly reduce the damage caused by second degree burns (776).

SOX9 helps strengthen and allow the body to create new bone (777), and carene, when administered in low amounts, also helps build new bone (778). While no studies have been done yet, it is highly probable that when carene is tested for SOX9 up-regulation it will show a positive result. Carene has been found in plants that have strong insect repellent properties (779).

Turkish folk medicine has used fruits from the Mediterranean cypress (Cupressus sempervirens) to treat common coughs and colds. The major substances in Mediterranean cypress are carene and α-pinene, with its fruits and leaves being abundant in tannins (780).

Lime Berry

The Lime Berry (Micromelum)
This bright orange edible berry is a little known fruit that contains abundant concentrations of carene (58.03%) and caryophyllene (55.41%) (781) and its essential oil contains 27.72% carene and 15.18% α-pinene (782). Also expect to see more studies on this berry in the future regarding its health and healing properties.

Bupleurum gibraltaricum contains 33.0% carene (783). When researchers induced swollen inflamed feet in mice, they found that after applying the essential oil of Bupleurum gibraltaricum, that it exhibited considerable anti-inflammatory activity. The researchers concluded this reduction in swelling was due to the carene (Pharmacological activity of the essential oil of Bupleurum gibraltaricum: Anti-inflammatory activity. M.A. Ocete et al. May 1989).

Further reading
Anti-aging Effect on Skin with the needles of red pine, Pinus densiflora. Seong-Kyu Park et al. Jan 2005.

Skin Care - Limonene
Out of 283 skin care products purchased from shops (784) in Lecce Italy, the following substances were found the most in the products (from highest to lowest) limonene 76.9%, linalool 64.6%, citronellol 34.1%, geraniol 31.5% and coumarin 30%. Limonene was most often found in rinse-off and make ups (70.7%) - sunscreen, lipstick, lip balm, shower gel and toothpastes, and linalool was found most often in skin care products that were left on the skin (87.7% - sunscreen, lipbalm and lipstick).

Human skin that had suffered from photoaging caused by UVB-induced photodamage was found to be significantly reduced when treated with limonene during a research study (785). The study concluded that limonene may exert dermato-protective effects on skin cells by activation of the NRF2-dependent antioxidant defense system and Limonene from citrus peel showed a powerful effect on tissue regeneration, significant anti-inflammatory effects, wound-healing effects and overall skin health. The authors conclude this may be due to the abundance of perillyl alcohol in the limonene (786). Perillyl alcohol is used as a mosquito repellent and for treating skin damage due to long-term sun exposure (787).

In conclusion, the 3 main monoterpenes found in Pine, Carene, Limonene and Pinene, show tremendous anti-aging potential and we expect to see more groundbreaking studies on these in the near future.

PRAL. Amino Acids and Bone Loss

PRAL stands for Potential Renal Acid Load. It is a concept developed by Manz and Remer (1994) showing an estimate for how well nutrients are absorbed. The less the PRAL, the more alkalizing the food is (higher PH). Foods that have a high number of acids that enter the bloodstream are pork, beef and poultry (787). Research is starting to show that diets with negative PRAL values are associated with reduced bone turnover. Studies by Shariati-Bafghi et al. (2014) found that PRAL levels were inversely associated in the bone mineral density of 151 postmenopausal women, even though the women had high intakes of calcium (above 800 mg daily) (788). There are some great charts available online that show foods that are acidic and foods that are alkaline using PRAL methodology. One such chart is **The PRAL Table. A Clinical Guide for assessing dietary acid levels** or you can just enter the search term **PRAL** into any search engine to find a list of related charts.

PDCAAS

PDCAAS is short for Protein Digestibility Corrected Amino Acid

Score. It is a composition of what amino acids your body requires by evaluating the number of amino acids in foods and then assigning them a specific score (789). For example, raw pistachio nuts show a PDCAAS score of 73. However this increases to 81 when the pistachios are roasted.

The Bone Restoration Properties of Pistachios

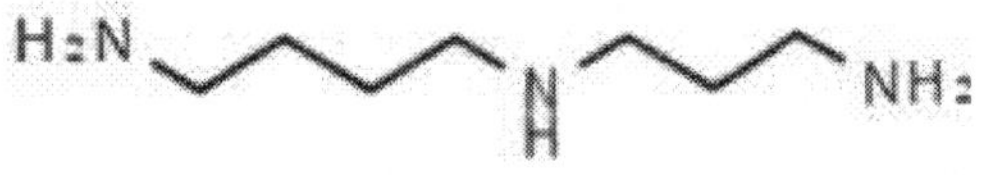

Spermine

Roasted and raw pistachio nuts contain an abundance of arginine, with leucine being the second most abundant (790). Pistachio nuts are also an excellent source of plant-derived protein (790) and they are also high in the anti-aging food spermine; as are almonds (791). Spermine has been found to extend lifespan when added to the drinking water of mice by strengthening their immune system (Zhang et al. 1997; Soda et al. 2005, 2009; Eisenberg et al. 2016).

How to Chelate Heavy Metals from the Body

One of the main symptoms of the gradual buildup of toxic metal buildup in the body is physical and mental fatigue. One popular method that has been used to remove unwanted toxic metals form the body with success is edible bentonite clay. More than 2,500 published research studies have documented the many beneficial effects of edible bentonite clay. It has been shown to remove cadmium in Nile tilapia fish, remove copper in sheep, remove cadmium from carp and more (792). Interestingly bentonite clay has been shown to increase the wool growth of sheep (792). Most important of all bentonite clay does not delete vital minerals and important vitamins from the bodies of humans and it may also help increase the activity of healthy gut flora activity which produce vitamins K and B as well as short-chain fatty acids (792). Bentonite clay is also being studied for its ability to reduce the growth of cancer cells and help flush out toxic pesticides from the body (792).

Lead Removal

Another metal chelator is glycine, which has been used to remove lead from bones in mice studies (793). Tuna contains variations in its levels of lead depending upon the age of fish, species, geographic region and size or age of the Tuna (794). When researchers visited the Laguna Region in Northern Mexico, known to have one of the highest levels of lead in the ocean out of all of Mexico, they discovered that an extract made of coriander was able to remove lead accumulation in rats (795) (796).

Cadmium Removal

The proteins in tomatoes have been shown to exhibit metal chelating properties when they encounter heavy metal ions (Tito et al., 2011; Steffens et al., 1986) and eating a tomato has been shown to significantly reduce the accumulation of heavy metals (Cd, Pb and Hg) in mice (Nwokocha et al., 2012). Thus tomatoes have been found to greatly reduce the levels of mercury, lead and cadmium in the bodies of mice (796). Curry leaves or grapes remove cadmium (796) and anthocyanins also remove Lead and Cadmium (796).

A natural herbal air pollution detox formula

A natural herbal formula that I used with success for many years while I lived in one of America's most polluted cities not only helped remove unnecessary toxic metals from my body, but also eliminated allergies that had been caused by the airborne pollution. This saved me tons of money in allergy medication. The herbal formula consists of the following: The formula is taken with 1.5 Tablespoons of Honey, 2 Vitamin C Capsules and 4 to 6 Cod Liver Oil Capsules.

1 teaspoon of garlic powder
1 teaspoon of echinacia powder
1 teaspoon of basil herb powder
2 1/2 teaspoons of brewer's yeast
1/2 teaspoon of yerba mate herb

1/2 teaspoon of parsley powder
1/2 teaspoon of clove powder
1/2 teaspoon of brazil nut powder
3/4 teaspoon of chili powder
3 bromelin enzymes
3/4 teaspoon of corriander
3/4 teaspoon of rosemary powder
A pinch of granulated lecithin (optional)
Between 2.0 and 10mg of vitamin B6 (optional)

Since inventing this formula over 20 years ago, a number of studies have confirmed the metal chelating abilities of many of the herbs shown above. For example, just as coriander leaves made into an extract have been used to reduce the contamination of heavy metals such as lead, mercury and copper that were found in shellfish and fish [(270)], when researchers [(797)] tested Yerba Mate leaves for their ability to remove lead and chromium from water, they found the Yerba Mate leaves exhibited good results, especially when the pH levels were below 5. The authors also stated that tannins and polyphenols also exhibited metal chelation activity.

The Misconception of Vitamin C supplements

Many people today believe that they have to constantly be taking Vitamin C in order to experience a strong and robust immune system. However the research shows that Vitamin C levels naturally rise in the body according to the right diet, and this includes certain foods. For example, people on Mediterranean Diets exhibit higher vitamin C levels circulating in their body and they also have higher levels of antioxidants and beta-carotene in their body [(798)]. Healthy levels of Vitamin C help enhance the absorption of iron in the body [(799)]. From my personal experience of using Vitamin C, I only find it of benefit when I feel a flu coming on or when I want to add synergy to specific herbal formulas, which I go into more detail in my anti-aging book series.

Anti-aging secrets of the Apricot

Dried apricots contain lutein, carotene and a significant amount of chlorogenic acid. They are also abundant in Gallic Acid and Quercetin (800). Apricot is compatible with many substances that have preservation properties. For example, when apricots are sprayed with salicylic acid for long term storage, the antioxidant levels in the apricots increase and when apricot is sprayed with ascorbic acid, chitosan, oligochitosan or potassium, it enhances their polyphenol levels (801) (802) (803). Adding just 1% of apricot fiber to yogurt increases the good probiotic bacterias Bulgaricus, L. acidophilus and Bifidobacterium (804) (805). Apricot peel is also abundant in polyphenols (806).

Using Coconut Oil for enhancing carotenoid bioavailability

Bovine bone marrow extract has been shown to contain a significantly high number of low molecular weight peptides and exhibit strong antioxidant activity (807). When coconut oil is added to bovine bile extract at 5%, it significantly enhances the bioavailability of the carotenoid zeaxanthin. Zeaxanthin is a pigment that improves eye health when combined with lutein and astaxanthin (808). The study concluded that mixing coconut oil with bovine bile extract enhances the absorption of zeaxanthin up to 42.8% (809).

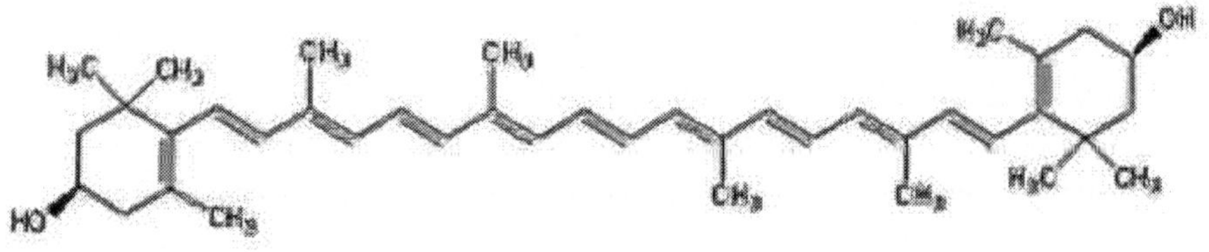

Zeaxanthin

Studies by Yuan et al. (2018) showed increases in lutein bioaccessibility between 20.9% and 46.5% when spinach puree was mixed with coconut oil and another research team discovered that the bioavailability of zeaxanthin from goji berries grew from 6.7%

to 13.3% simply by adding just 1% coconut oil (Hempel et al., 2017) (810).

Easy to grow at home anti-aging plants

The main advantage of growing your own anti-aging plants is they can be organic, have no contamination, and best of all they can be harvested and eaten fresh, not to mention the cost and time saved having to go to the store or order them online. Because all parts of the plant are edible, you end up using less and you can make the roots and leaves into a tea, which allows you to extract the anti-aging substances without the loss of anti-aging nutrients that occur during processing, handling and storage. You can also make an extract from the plant. See my book titled: The Official Guidebook of How to Make Tinctures and Alchemy Spagyric Formulas, for instructions on making your own herbal extracts. Easy to grow at home anti-aging plants include the following:

Angelica keiskei Koidzumi (ashitaba). The main anti aging substance present in Ashitaba is **Dimethoxychalcone**, which has been scientifically proven to extend lifespan up to 20% or more (811). When researchers (812) in Europe screened 180 subclasses of flavonoids for their anti-aging effects, they discovered that **dimethoxychalcone outperformed reserveratrol**, polyphenols, specific flavonoids and yielded similar protective effects as the pharmaceutical Rapamycin (also a powerful anti-aging substance). It also **prolonged the lifespan across multiple species**, including worms, human cells, yeasts and fruit flies. The researchers also discovered it exerted protective effects on the livers and hearts of mice. When Dimethoxychalcone was added to neuronal tissues in mice, it was found to ameliorate Parkinson's Disease phenotypes (813). Dimethoxychalcone is also found in the following plants that are also edible and are also easily grown at home (814):

Three-lobe Beggarticks (Bidens tripartitus). This plant is a part of the sunflower species. Its cooked leaves are edible.

Pongame oiltree (Pongamia pinnata). The fruits and flowers of these trees are edible, however the seeds are poisonous.
Tickseed flowers (Coreopsis lanceolate flowers). Flowers are edible.
Mexican Kidneywood (Eysenhardtia polystachya). All parts are edible.
The inner bark of **Heartwood bark** is edible. The bark can be made into an extract, and it has been used for treating breast cancer.

Another anti-aging plant is the Hawaiian plant known as **Cobbler's Pegs** (**Bidens pilosa L**), which was mentioned earlier where it was used for treating cataracts. All parts of the plant are edible. The leaves have a resinous flavor and are eaten raw, used in stews, or dried for long term storage (815). This plant has been used as feed for animals, showing no adverse side effects when eaten in bulk (815). An extract of Cobbler's Pegs was found to prolong the lifespan of mice and it also exhibited significantly improved neuronal brain functioning (816). Expect to see many more studies on this important plant in the future.

The STW 5 Herbal Formula

STW 5 is an all natural herbal extract that came into the limelight over 50 years ago. Since then it has been estimated to have been consumed by over 25 million people for a variety of conditions (817). In some cultures there exists the perception that natural herbal remedies are moderately effective. However over the past decade there are hundreds, if not thousands of studies proving this theory invalid (817). Published studies over the last few decades have confirmed the effectiveness of STW 5 (817). STW 5 is currently sold under the brand name **Iberogast** and is used for treating gastrointestinal disorders of the lower and upper abdomen, abdominal cramps, diarrhea, heartburn, indigestion, bloating, gas and constipation.

STW 5 is made into an extract with alcohol with the following herbs: Peppermint Leaves (*Menthae piperitae folium*),

Candytuft (*Iberis amara totalis recens*), Wild celery (*Angelicae radix*), Liquorice Root (*Liquiritiae radix*), Milk thistle fruits (*Cardui mariae fructus*), Greater Celandine (*Chelidonii herba*), Chamomile (*Matricariae flos*), Lemon Balm (*Melissae folium*) and Caraway fruit (*Carvi fructus*).

STW 5 has proven its effectiveness against a placebo and the pharmaceutical prescription drug Cisapride, showing superior effects compared to the Prescription Drug Cisapride (817). In 3 studies (818) involving participants diagnosed with dyspepsia (indigestion), which compared STW 5 to a placebo, the results found statistically clinical superiority of STW 5 compared to the placebo. When STW 5 was used to treat irritable bowel syndrome, it reduced IBS levels by 1.5 points more than the placebo, which was also a significant clinically relevant effect ($p < 0.0004$), and all with no serious adverse side events (818). In a large scale non-interventional study involving 980 children aged between 3 and 14 that had been diagnosed with gastrointestinal disorders who took STW 5 for a week, they experienced a decrease in their symptoms by 76% (818).

STW 5 is very similar and offers very similar effects to the St. Germain Formula, which has been in use longer than the STW 5 formula (estimated to be in use for over 200 years). The St. Germain Formula offers powerful flu resistance, constipation relief, enhances lung strength and improves bone strength. The formula contains the following herbs:

3 parts senna
1.5 parts elderberry
1 part fennel seed (after crushing into a fine powder, slightly more is added to the final formula)
0.5 part star ansie

The Senna is ground up than brought to a low heat so it does not burn. This reduces the size of the Senna. The powdered Senna is than placed in a pot over a flame with low heat. Next a fine mist sprayer with cream of Tartar powder and water mixed in it is sprayed on the warm Senna until the Senna turns a very light yellowish brown. After the Senna is dried, it is removed from the pot and crushed into a fine powder using a mortar and pestle. Next the other herbs are measured out and mixed in with it. It is than stored in an air-tight container and made into a tea or made into an extract. See my book titled: The Official Guidebook of How to Make Tinctures and Alchemy Spagyric Formulas for more details on published studies regarding effectiveness of the St. Germain extract and how to best use it.

Diketopiperazines

Diketopiperazines act to suppress appetite and are the smallest known class of cyclic peptides. They are found in the body's cerebrospinal fluid and gastrointestinal tract. Diketopiperazines are also found in sherry, Shaoxing wine, soy sauce, barley tea, roasted malt, Korean kimchi, sweet cooking rice wine, bitter potato and vinegar. Diketopiperazines are also found not just in roasted coffee beans, but also coffee. Diketopiperazines can be created in the lab by using high-pressure, high temperature processing of plant-based peptides (819). Diketopiperazines are found in abundance in Asparagus and have also been found to reduce heat shock proteins (820) (821). A commercially prepared substance that was made from the stems of Asparagus called ETAS® was found to significantly reduce heat shock proteins, due to its abundance of diketopiperazines (822). Asparagus has also been used to detox indoor air of the pollutant sulfur dioxide (823).

Cyclodextrins

Cyclodextrins have been shown to act as "*secondary antioxidants*" enhancing the ability of traditional antioxidants. Cyclodextrins allow for enhanced water-solubility of specific substances

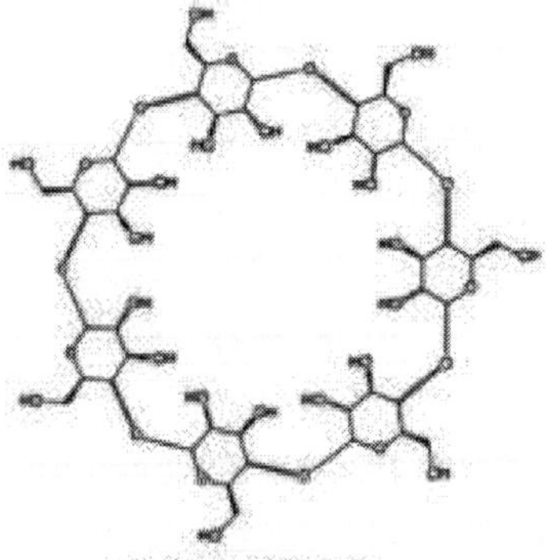

Cyclodextrin

that may be hard to dissolve in water. This allows for increased bioavailbilty when added to substances that show low bioavailability. Adding cyclodextrins to mandarin juices enriched with pomegranate extract and goji berries protects retinol, vitamin C levels and extends antioxidant levels for up to a month (825).

Guaiacol is a powerful anti-fungal agent and is also used as a food preservative (826). When Guaiacol is combined with cyclodextrin, it has been found to help stop infection occurring after a tooth has been pulled and to also help form new bone (827).

The Bone Strengthening Properties of Melon

When researchers (828) examined the effects of extracts made from muskmelon (Cucumis melo L) and its effects upon mineralization and bone strength in mice with osteoporosis, they discovered the extract exhibited the ability to protect against bone loss in rats that had osteoporosis. They also discovered that main phytochemical present in the melon was caffeic acid and that the main polyphenols were ellagic acid, catechin, gallic acid, ferulic acid, kaempferol and chlorogenic acids, which were present in extracts of the melon peels and seeds. The study (828) also stated that when the melon was heated to between 140°C and 150°C, that the polyphenol and antioxidant levels rose, with the best results obtained when heated to 150 °C, which increased these levels up to 25 times or more compared to the unheated samples. The authors also state that when Citrus peel extracts are heated, they show significant increases in antioxidants.

Wines and grapes abundant in resveratrol

Resveratrol exists in grapes of the Vitis labrusca and Vitis vinifera variety (829). Dark-skin grapes of the muscadine variety contain above average levels of resveratrol, compared to the bronze-skinned variety [829]. The muscadine grape seeds also have higher concentrations of resveratrol than other parts of the grape with higher levels in the dark skinned grapes [829]. V. vinifera and V. labrusca wines made from the Muscadine variety of grapes contain good levels of resveratrol [829]. Other varieties of wines that contain an abundance of resveratrol include Israeli Merlot and Shiraz red wines [830]. Above average levels of trans-resveratrol are found in Merlot wines grown in the Golan vineyards in Israel, compared to other regions. [830] Studies suggest the high relative humidity levels (up to 89%), especially during springtime, are creating more trans-resveratrol in the grapes. These wines also had higher gallic acid and quercetin amounts [830]. Other vineyards that have high concentrations of trans-resveratrol are in vineyards on the relatively heavy brown Mediterranean soils (basalt weathering product). These are soils similar to the Golan Heights. The soils are created via the weathering of tuffs and basalts and also exhibit high levels of clay in the soil. However samples taken from the vineyards located on Rendzinas soil (sloping karst and mountain landscapes) and Terra Rossas soil (red, well-drained, clay to silty soil that has neutral pH conditions), which exist among limestone and dolomite outcrops, and in the Samaria, Galilee and Judea regions, do not have significant levels of trans-resveratrol in their grapevines [830]. Wines richest in trans-resveratrol include St. Laurent and Pinot noir red wines (Contribution of Red Wine Consumption to Human Health Protection. Lukas Snopek et al. Jul 2018).

Resveratrol has a high hydrophobicity value. Hydrophobicity is when dust and dirt lose their ability to cling to a surface [831]. Cyclodextrins have been used to protect reserveratrol against decay [832]. This may be because Cyclodextrins have a structure that is able to deeply penetrate reserveratrol and CQ10. Combining cyclodextrins with CQ10 significantly increases the bioavailability of CQ10 [833].

Is Grape Juice better than Red Wine?

A research study (834) sought to find out which had more antioxidants, red wine or red grapes. The study found that antioxidant levels were higher in red grape juice. The study stated that this may be due to the enhanced intestinal absorption of the anthocyanins due to possible synergistic activity occurring from the glucose in the red grape juice. Red grape juice also has more resveratrol and anthocyanins compared to red wine. Grape juice contains the sugars glucose and fructose (McLellan and Race, 1990) and glucose is one of the ingredients in the longevity Mushroom called Maitaike.

Is Piceatannol better than resveratrol?

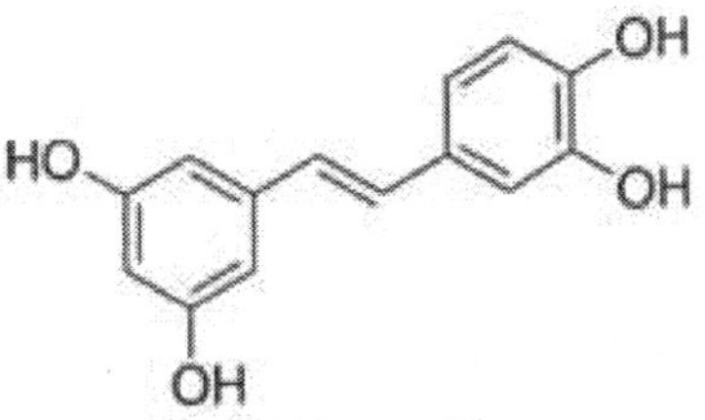

Piceatannol

The polyphenol Piceatannol is the hydroxyl compound of resveratrol and has shown stronger cytoprotective and antioxidant potential than resveratrol (Piceatannol Is Superior to Resveratrol. In Sil Park et al. Feb 2021). It is found in grape skin and occurs in abundance in passion fruit. It is currently being explored for treating atherosclerosis and hypercholesterolemia. When researchers fed worms Piceatannol, they lived 20% longer as well as exhibited enhanced resistance to stress (Piceatannol extends the lifespan of Caenorhabditis elegans via DAF-16. Peiyi Shen et al. May 2017).

Beetjuice

A study by USC (University of Sunshine Coast Australia), stated that beetroot juice promotes healthy blood flow, reduces arterial stiffness and reduces nitrate absorption in the body up to 9 fold (Beetroot juice a bloody good boost for arteries. UniSC News. 8 May 2019).

Peptides

Peptides are made from plants by heating them to 100 C or more for between 2 and 7 hours using a pressure-resistant extracting apparatus. The most common peptides are simply composed of two-amino acids. This simple structure allows for easy bioavailability into the body. Diketopiperazines and Cyclodipeptides are also peptides. Foods and substances that contain these are some of the most powerful anti-aging foods. The antifungal that is produced by the bacteria B. cereus is classified as a diketopiperazine. B. cereus is a good bacteria found in soil, rennet and ricotta cheeses, beans, rice, potatoes, peas and spices. It is especially abundant in cumin powder and black pepper. The majority of dried spices contain B. cereus. These include fennel leaves, basil, thyme, oregano, rosemary, coriander, lovage, allspice, marjoram, bay leaves, tarragon, parsley leaves, and the herbal blend Herbs de Provence, with the highest levels found in thyme (835).

Recently discovered substances that extend lifespan -
A 75% ethanol extract of Japanese honeysuckle (L. japonica) was found to extend lifespan in worms up to 21.87%, and showed greater life-prolonging effects than dicaffeoylquinic acid or chlorogenic acid (Yang et al. 2018). The authors of the study also looked at Ginseng Oil and stated that both panaxynol and panaxydol, which are both present in Ginseng Oil may exhibit synergistic effects regarding their anti-aging effects (836). The stems of the Japanese honeysuckle contain 31% hexadecanoic acid (also found in celery seeds) with the flowers containing 11% linalool (Chemical composition of the essential oils from the flower, leaf and stem of Lonicera japonica. Nenad Vukovic et al. May 2012).

United States Patent #6190652 titled: Volatiles of Japanese honeysuckle flowers as attractants for adult Lepidopteran insects, stated that Japanese honeysuckle contains Carene.

Green Tea Significantly Extends Lifespan
A published study (837) stated that fruit flies fed green tea extract exhibited a mean lifespan of 16% and when mice drank the green

tea they lived far longer than the mice who didn't drink the green tea. When EGCG, a substance found in abundance in Green Tea, was given to mice it significantly extended their life span up to 2 years. The study also stated that the EGCG delayed death in mice that were healthy by reducing their age-related inflammation, oxidative stress and kidney and liver damage. It also stated the significant life span extending qualities attributed to EGCG in the worms may be attributed to the free radical scavenging effect and enhanced levels of Superoxide Dismutase (SOD).

When worms were subjected to high temperature stress (35°C), their average life expectancy increased up to 13.1% with EGCG **increasing their average lifespan up to 172.9%** [837]. The author of the study [837] also stated that the lifespan extending effects of EGCG may be attributed to its ability to up-regulate stress-resistance related proteins (heat shock proteins and superoxide dismutase-3) and reduce free radicals.

According to the peer reviewed study: A literature review of flavonoids and lifespan in model organisms, that was published by Cambridge University Press on 9th September 2016, and written by Kathrin Pallauf and colleagues, these are the flavonoids that extended lifespan the most in animals. It is interesting to note that Baicalein, which was the most powerful, that only a small amount was necessary for the best results:

- Baicalein extended lifespan 45% at a dose of 100um
- Quercetin 3-0-glucoside extended lifespan 39% at a dose of 200um
- Myricetin extended lifespan 33% at a dose of 100um
- Quercetin (from onion species Kito Momiji) extended lifespan 20% at a dose of between 50 um and 250um
- Silymarin extended lifespan 25% at a dose of 50um
- Flavonoids from onion (from onion species Kito Momiji) extended lifespan 20% at a dose of 20um
- Epigallocatechin gallate extended lifespan 20% at a dose of 100um

Terpenoids from Rosemary (*R. officinalis*) extended lifespan 26% (838). The molecular symbol for carnosic acid, which is found in Rosemary, is shown on the right. Ferulsinaic acid (*commonly confused with Ferulic acid*) from the Ferula species of plants was shown to not only protect against AGE's, but also offered significant protection against ROS, heat stress and oxidative stress, extending lifespan +42% (839). Pomegranate extends lifespan 36% (840). What's most interesting is that the most abundant substances in Spanish Pomegranate peels are carene, terpineol with minor amounts of α-pinene, β-pinene, limonene and β-myrcene (841).

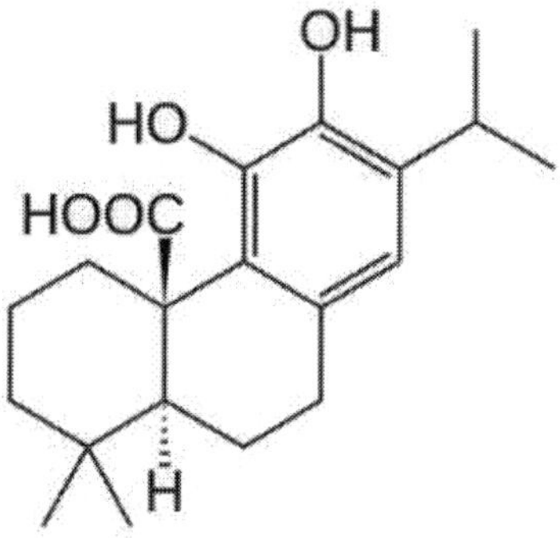

Carnosic acid

An extract made from Alpinia zerumbet Leaf (Shell Ginger) extended the life of worms up to 22.6% by reduicing heat shock. These results were better than the lifespan extending substances resercveratrol or quercetin (842).

Anti-aging Synergy

In some cases, taking an anti-aging extract with the herb that has been made into a capsule or supplement adds synergy to its effects. Examples include taking Goji berries with Goji berry tea, Roobios tea with 5 drops of Roobios extract added to the tea or capsules of Pycnogenol with Pine Needle Tea.

The anti-aging power of the Caper

Out of all the many anti-aging foods, Capers, with its above average quercetin content, also contains salicylic acid, quinic acid, chlorogenic, ferulic acid, rosmarinic acid, vanillic, caffeic, syringic, p-coumaric and vanillin (843) (844). This makes it a great food to add to your favorite dish if you want to increase the bioavailability of healthy nutrients. Other foods abundant in quercetin include:

lovage (845), dark chocolate, cloves, black elderberries, mexican oregano, onions and red wine (846).

The Interesting Chemical Reaction between Bacon and Wood Chips

Cooking bacon in woodchips produces the following substances: dimethyl trisulfide, methyl ester, hexanal, guaiacol and furfuryl alcohol (847). It is my hypothesis that the fat in the bacon acts as a catalyst for a long, slow process of the Maillard reaction to take place, creating these anti-aging substances. Let's take a look at these anti-aging substances in detail.

Dimethyl trisulfide has been found to be an effective cyanide antidote in mice studies (848). Methyl ester is found in rosemary (849) and methyl ester is used to treat cervical cancer which is resistant to the chemotherapy medication cisplatin (850). Hexanal is used to increase the shelf-life and quality of stored fruits (851) and when pork is roasted, the Maillard reaction produces mono-MGO propyl gallate (852), which exhibits antioxidant effects (853).

What is NRF2?

NRF2 is a protein found in your body which helps regulate antioxidant proteins to protect cells against oxidative damage by reducing inflammation and fighting free radicals (854). Low NRF2 activity has been shown to suppress or reduce SOX9 activity (855). Animal studies have shown that even mild activation of NRF2 extends lifespan due to its ability to affect ROS and attenuate aging-related traits (856).

What natural herb increases ATP Levels?

ATP is short for adenosine triphosphate, which is a substance found in your cells whose molecule provides energy to your body by acting as a fuel source for cellular processes. Coriander has been shown to significantly increase adenosine triphosphate (ATP) levels in mice that underwent physical exercise (857). Coriander also protects the skin against photoaging (858).

Using Magnetic Fields to Destroy Bad Bacteria

A June 2022 study (859) used rotating magnetic fields (between 25 Hz and 50 Hz) on pine essential oil. After two and a half hours, the antibacterial properties of the pine oil increased. The pine oil consisted of 14.17% carene, 11.58% limonene, 28.58% α-pinene and 17.79% β-pinene.

Polyethylene Glycol (PEG)

While Polyethylene Glycol gets a bad rap sometimes, when properly used, it has been shown to have many benefits. For example Polyethylene Glycol has been shown to improve functioning in Synaptosomal Preparations following Spinal Cord Injury (860) and another study found that when Polyethylene Glycol was added to the diet of goats, it resulted in an increase in their milk production (861). The bioavailability of Polyethylene Glycol becomes enhanced when combined with magnesium, and when it does so, it creates enhanced neuroprotection (862). Food grade PEG is a common food additive and can be readily bought online.

The Dangers of Glucose Rich Diets

A diet rich in glucose has been shown to increase the levels of palmitic and myristic acids, which has been shown to decrease the lifespan of worms. However this was restored when the worms were fed arachidonic and linoleic acids (863). Linoleic acid cannot be produced by the body and must be obtained from foods such as nuts and seeds. Arachidonic fatty acid is found in egg yolks (864).

Lowering Blood Glucose Levels

Out of 30 medicinal plants used in folk medicine (865), Ayurvedic, Unani and Siddha treatment methods were found to be the best for lowering blood glucose levels in mice. The plants were: Chiretta (*Swertia chirayita*), Bitter Melon (*Momordica charantia*), Cluster fig tree (*Ficus glomerata*), Greek hayseed (*T. foenum-graecum*), Drumstick tree (*Moringa oleifera*), Black Plum (*Eugenia jambolana*), Guduchi (*Tinospora cordifolia*), The Banyan (*Ficus benghalensis*),

Velvet Bean (*Mucuna prurita*), Myrobalan (*Terminalia bellirica*), Madagascar Periwinkle (*Vinca rosea*), Premna integrifolia, Egyptian Pea (*Sesbenia aegyptiaca*), Neem Tree (*Azadirachta indica*), Bael Fruit (*Aegle marmelos*), Indian Bay Leaf (*Cinnamomum tamala*), Hamilton's bamboo (*Dendrocalamus hamiltonii*), Ginger (*Zingiber officinale*), Snake Gourd (*Trichosanthes cucumerina*) and Holy basil (*Ocimum sanctum*).

Reduction of Heat Shock Proteins to Extend Lifespan

If there is any key to solving the aging mystery, it is proteins. Proteins are a double edged sword. They can rejuvenate the body as in the case of plant proteins, or in the case of excessive meat consumption, they can contribute to ill health of the body; especially if cooked at high temperatures for long periods of time. Heat shock proteins are a class of proteins created by your cells in response to stressful conditions. Examples include cold, heat shock, UV light or wound healing.

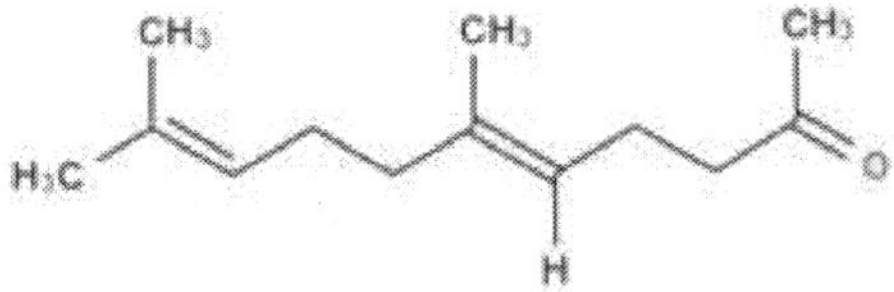

Geranylacetone

A research study (866) that screened 80 substances for their ability to reduce heat shock proteins found that the substance Shikonin was the most potent in its ability to reduce heat shock proteins. Shikonin is a naturally occurring naphthoquinone which exists in the dried roots of Purple Gromwell (Lithospermum erythrorhizon) as well as in the herb Comfrey. Geranylacetones are produced in plants as an insect repellent and to deter pests (867). Geranylacetone is used to reduce the impact of heat shock proteins that occur in the brain and lungs (868) (869). Geranylacetone is found in Scutellaria baicalensis, Sweet Cherry and Hornwort (Ceratophyllum demersum) (870). Shikonin is

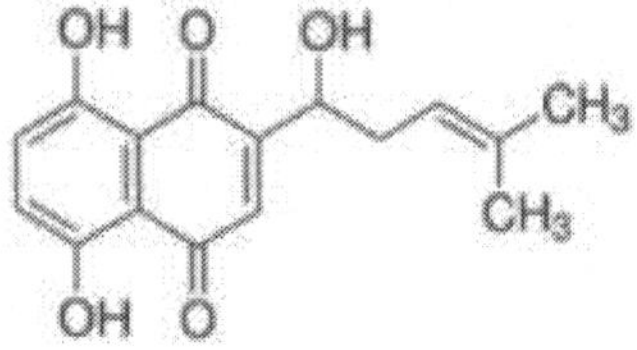

Shikonin

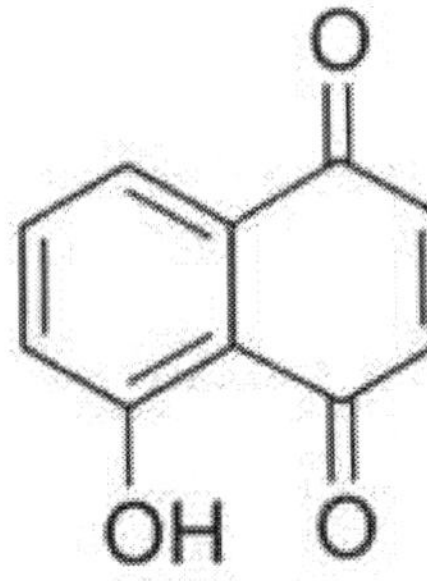

Naphthoquinone

also found in Common Bugloss (Anchusa), Goldendrop (Onosma), Common Gromwell (Lithospermum), Spanish bugloss (Alkanna), Arnebia and Blueweed (Echium) (871).

Naphthoquinones, which repel insect herbivores, when given in low doses, extend lifespan by reducing oxidative stress (872).

Supplement Quality

When you buy a herbal supplement, or even some foods and beverages, know that the quality of the supplement varies due to the herbs being harvested according to season and geographic location. For example, the Anthocyanin levels in wine are significantly less when there has been a particularly above average wet year. This occurs because an increase in the water content of foods that contain Anthocyanins dilutes their content (Esteban et al. 2001, Kennedy et al, 2002) (873).

Instantaneous Healings

If you are a fan of miracle healings, www.**realnewsrevealed**.com has numerous videos from people from all walks of life demonstrating remarkable healings on video, and if you are interested in published studies showing the effects of herbs on aged mice, I highly recommend the research paper: **Effect Anti-Aging Effects of Some Selected Iranian Folk Medicinal Herbs-Biochemical Evidences**. Azadeh Mohammadirad et al. Nov 2013.

Aging and brainwave frequencies

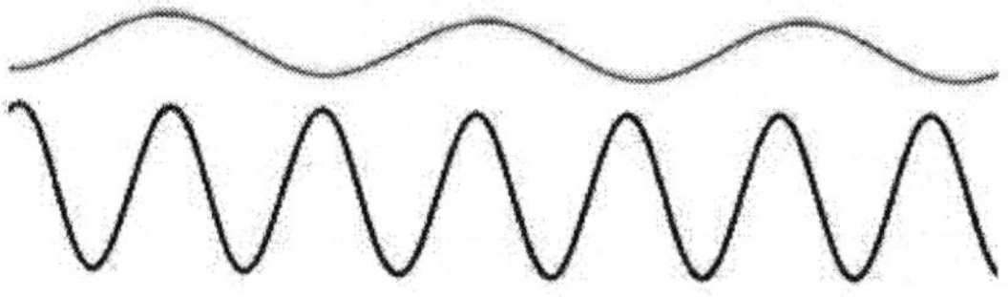

When a survey looked at brainwave frequencies over the course of a human lifetime, alpha brainwaves were strongest between 11 and 12 years of age, than began to decrease until age 40 (874). In a research study of twenty volunteers, the inhalation of the essentials oils patchouli (especially patchouli) or ylang ylang was shown to significantly enhance alpha brain waves (875). Patchouli has been shown to express significant anti-inflammatory activity (Anti-inflammatory effect of patchouli alcohol. Yan-Fang Xian et al. May 2011).

Essential oils that enhance alpha brain waves when inhaled include: **Peppermint, Lavender, Champaka, Eucalyptus and Zizyphus jujuba oils** (876). Alpha activity has been shown to synchronize with one's heartbeat while practicing the Heartmath Technique, which is a simple technique that takes only a few minutes where you imagine yourself breathing in and out through your heart. You can learn more about Heartmath by visiting www.**heartmath**.org (877) (878). What is most interesting is when a set of volunteers experienced theta entrainment, they exhibited increases in their memory (879).

Wine Bioavailability

A research study (880) looked at what substances in wine were absorbed by the body the most. The study found that phenolic acids (*which are simplified Polyphenols*) were absorbed the most, while flavonols were the least absorbed (880). Other research has shown that when worms were given phenolic acids, that it significantly increased their heat-stress resistance and lifespan and that all their longevity pathways became activated by the phenolic acids (881). Getting back to the original wine bioavailability study (880), gallic acid bioavailability in the Cabernet red wines were highest (24.01%) and Chinese red wine from the Helanshan mountains showed the most antioxidants. Previous research has shown that the bioavailability of anthocyanins absorbed by the body is low, while the phenolic acids, including ethyl gallate and other small molecular weight phenolic substances, are more easily absorbed

Further Reading

Identification of the 100 richest dietary sources of polyphenols: an application of the Phenol-Explorer database. J Pérez-Jiménez et al. Nov 2010.

Aged Garlic Strengthens Memory

Garlic has been used to generate pink-red pigments from natural amino acids (882) and Garlic has been used to make gold (883). Aged garlic has been suggested for treating age-related memory deficits and physiological aging in humans (884). A resesarch study examining 92,505 people found that people who ate garlic more than 5 times a week lived longer than people who ate garlic less than once week (885). After 3 weeks of aging, fresh garlic exhibits increases in S-1-propenyl-l-cysteine sulfoxide (886) and animals who ate S-1-propenyl-l-cysteine sulfoxide, which is present in aged garlic extract, showed that its bioavailability was up to 100% (887). Aged garlic has been shown to enhance memory and learning (888) and Garlic essential oil contains carene and limonene (889).

Super Bioavailable Nutrient Meal

This meal is a great way to allow your body to absorb numerous nutrients. You will need chickpeas, 1/2 tomato, olive oil, capers and smoked oysters.

Instructions:

1. Soak chickpeas for 4 to 6 hours to remove the phytic acids.
2. Next strain out the water from the chickpeas into a small pot and bring to a boil for 1 to 3 minutes, placing the Chickpeas aside.
3. Once the water starts to boil add the capers.
4. Next reduce the temperature of the boiling water and add the liquid from the beets, and allow it to soak for 1 to 2 minutes.
5. Boiling the water removes the bitterness and tart taste of the capers and chickpeas.
6. Slice up 1/2 tomato and smother with olive oil.
7. Strain out the water and add 2 cans of smoked oysters, sliced tomato, olive oil and beetroot.

Food Sources Of the 6 main Anthocyanins	
Delphinidin	Red Wine & Berries
Petunidin	Black Goji Berries And Purple Potatoes
Malvidin	Blueberries
Cyanidin	Cherries
Peonidin	Berries
Pelargonidin	Stawberries

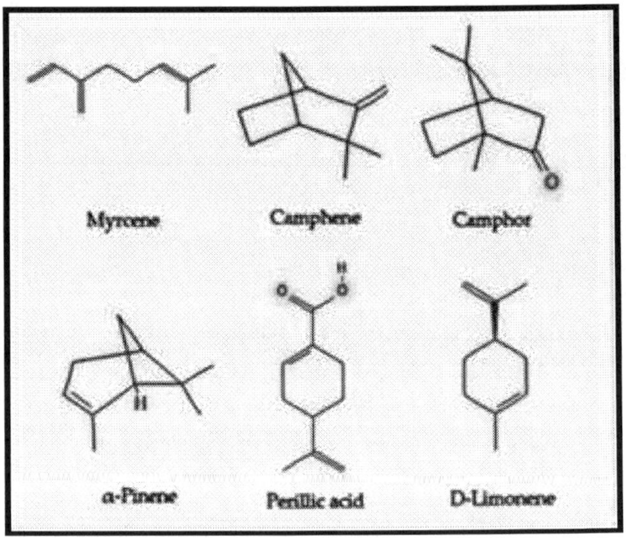

Above is a list of monoterpenes proven to delay age-related diseases as well as increase lifespan (890). The chart on the following page lists many of the substances mentioned in this book showing the major compounds in each plant or herb.

The Dominant anti-aging substances in common foods, plants and herbs	
vanilla	vanillin, ethyl vanillin, coumaric acid, ferulic acidvanillyl alcohol and vanillic acid.
turmeric	curcumin.
thyme	cymene, gamma-terpinene, linalool, borneol, thymol and carvacrol.
star anise	anethole, limonene, estragole and phenylpropanoids.
sesame seeds	sesamin, tocopherol, stigmasterol, phytic acid, linoleic acid, oleic acid, β-sitosterol, campesterol and stearic acids.
saffron	camphor, thujone, β-thujone, viridiflorol and borneol cineole.
rosemary	ursolic acid, camphor, limonene, camphene, borneol, cineole, linalool oxide, carnosol, rosmarinic acid, carnosic acid, α-pinene and bornyl acetate.
red peppers	β-carotene, zeaxanthin, lutein, capsanthin, capsaicin and caffeic acid.
parsley	apiole, apigenin, myrcene, rutin, myristicin, p-1,3,8-menthatriene and β-phellandrene.
onion	quercetin, allyl propyl disulphide and protocatechuic acids.
mint	limonene, carvone and 1, 8-cineole.
ginger	gingerol, paradol, bisabolene, α-farnesene, β-phellandrene, shogoal, zingiberene, citral (neral and geranial), cineole and zingerone.
fennel	estragole, sabinene, β-myrcene, trans-anethole, fenchone, limonene, anisaldehyde, α-pinene, β-pinene and camphene.
	α-pinene, limonene, dill ether, sabinene, n-

dill	tetracosane, neophytadiene, α-phellandrene,n-heneicosane, n-docosane, n-tricosane, n-nonadecane, n-eicosane, β-myrcene and α-tujene.
coriander	linoleic acid, oleic acid, petroselinic acid, palmitic acid,vaccenic acid, myristic acid and stearic acids.
cinnamon,	cinnamaldehyde, linalool, humulene, cinnamyl acetate, cineole, eugenol, coumarin, τ-cadinol, ethyl cinnamate and β-caryophyllene.
celery seeds	lupeol acetate, hexadecanoic acid, 2 isopropyl-5-methyl-phenol, octadecanoic acid, stigmasta-5,22-dien-3β-ol, lup-20(29)-en-3-yl acetate and (3β, 24s)-stigmast-5-en-3-ol.
cardamom	limonene, linalool, terpinolene, 1,8-cineole, α-terpinyl acetate, myrcene and linalyl acetate.
black pepper	α-pinene, β-pinene, α-phellandrene, piperine, β-caryophyllene, myrcene, terpinolene, limonene and carene.
black cumin (*nigella sativa*)	β-pinene, p-cymene, p-mentha-1,3-diene-7-al, thymoquinone, cuminaldehyde, γ-terpinene and p-mentha-1,4-dien-7-al.
bay leaves	α-pinene, limonene, alpha-terpinyl acetate, 1,8-cineole and terpinene-4-ol.
basil	linalool, 1 8-cineole, estragole, eugenol, caryophyllene, β-ocimene, methyl cinnamate, α-cubebene and α-farnesene.
Asafoetida (*also called Asafetida*)	**rhamnose**, glucuronic acid, **ferulic acid**, umbel-liferone, asaresinotannols, galactose, l-arabinose, 2-butyl propenyl disulfide, farnesiferols a, b, c and glucose.

anise	anethole, para-anisaldehyde, methyl cavicol, estragole, γ-and hymachalen.
Chronic diseases, inflammation, and spices: how are they linked? ajaikumar b. kunnumakkara, et al. jan 2018.	

Food sources of Important Vitamins		
Nm	**Source**	**RDA**
A	**Carotenoids:** green leafy vegetables (spinach, broccoli), carrots, orange flesh fruits (mangoes, melons and persimmons) and orange-flesh sweet potatoes. Red palm oil and pumpkin. **Retinol:** cheese, fish, butter, egg yolk and milk.	2,333-3,000 IU / 700-900 µg
D	**Foods:** mushrooms, liver, beef, egg, yolk, fortified cereal, dairy products, veal and fatty fish. **Sunshine**: ultraviolet B radiation.	600–800 IU / 15–20 µg
E	**Edible vegetable oils:** avocado, sunflower seeds, fruits (kiwifruit, mango), leafy green vegetables (chard, spinach), nuts (almonds, peanuts) and nut spreads.	22–33 IU / 15 µg
K	**Phylloquinone:** broccoli, spinach, cabbage, vegetable oils (olive, soybean, canola). parsley, and collard greens. **Menaquinone:** Fermented cheese and curds, Natto (fermented soybeans),	90–120 µg
C	Citrus fruits, tomatoes, brussels sprouts, broccoli, lettuce and potatoes.	75–90 mg
B1	**Citrus fruits:** pineapple, orange juice, strawberries and grape. Fish, legumes (beans	1.1–1.2 mg

	and lentils), asparagus, whole grain cereals, squash, nuts, brewer's yeast and soymilk.	
B2	Brewer's yeast, nuts, dark green leafy vegetables, whole grains cereals, eggs, milk and yoghurt.	1.1–1.3 mg
B3	Yogurt, cheese, legumes, milk, egg, meat, fish, fruits (avocados, dates, fogs and prunes), mushrooms and nuts.	14–16 mg
B5	Red fish, cereals, brewer's yeast, legumes, tomatoes, potatoes, egg yolk, milk, green leafy vegetables broccoli and mushrooms.	5 mg
B6	Milk, yogurt, mushrooms, broccoli, fish, shellfish and chicken.	1.3–1.7 mg
B7	Chickpeas, maize and whole grain cereals, starchy vegetables, bananas, potatoes, fish (tuna, salmon), beef, liver, nuts (peanut, walnut) and fruits.	30 µg
B9	**Nuts:** peanuts, walnuts, cereals, yeast, asparagus, milk, lentils, egg yolks, wheat germ and beans.	400 µg
B12	**Fish:** trout, salmon, herring and sardines, shellfish, spinach, dark leafy greens, asparagus, beets, turnips, mustard greens, milk and milk products.	2.4 µg

Revisiting food-sourced vitamins for consumer diet and health needs: a perspective review, from vitamin classification, metabolic functions, absorption, utilization, to balancing nutritional requirements. Chigozie E. Ofoedu et al. Sept 2021

Average Vitamin Levels in Common Foods		
Vitamin	**Source**	**Vitamin Contents**
A	Chicken liver Milk Cheddar cheese Beef liver	308 µg 6.2 µg 7.1 µg 679 µg
D	Vitamin-D fortified non-fat milk Eggs Oiled Sardines	3.10 µg 1.03 µg 1.15 µg
E	Sunflower Seeds Peanuts Sunflower oil Almonds	7.4 mg 2.2 mg 5.6 mg 6.8 mg
K	Cabbage Broccoli Spinach	34 µg 160 µg 27 µg
B1	Breakfast cereal (fortified) Enriched white rice Cooked Tuna	1.5 mg 1.4 mg 0.2 mg
B2	Scrambled Eggs Fortified Instant Oatmeal Fat free plain yoghurt	0.2 mg 1.1 mg 0.6 mg
B3	Whole wheat bread Fortified Cereal Tuna Turkey	1.3 mg 20 mg 8.6 mg 10 mg
B5	Avocado Plain non-fat Yoghurt Fish trout	1 mg 1.6 1.9 mg
B6	Chickpeas Tuna	1.1 mg 0.9 mg

	Potatoes	0.4 mg
B7	Sweet Potato	2.4 µg
B9	Spinach Broccoli Asparagus	131 µg 45 µg 85 µg
B12	Salmon Low fat milk Fortified breakfast cereal Canned tuna	4.8 µg 1.2 µg 1 .5 µg 2.5 µg
C	Tomato Romine lettuce Orange Juice Grape Juice	17 mg 28 mg 93 mg 70 mg
Revisiting food-sourced vitamins for consumer diet and health needs: a perspective review, from vitamin classification, metabolic functions, absorption, utilization, to balancing nutritional requirements. Chigozie E. Ofoedu et al. Sept 2021.		

Additional combinations that exhibit bioavailable synergy include: Sardines with cottage cheese, Spirulina with Yogurt, Poppy seeds with fried eggs and Banannas dipped in Teff (*Eragrostis tef*)

Exploring Synergistic and Longevity Substances in Foods

The tables on the following pages show which foods and substances extend lifespan, as well as synergistic substances that go with them. **Note**: Even through a substance in the following charts may show that it does not activate SOX9, or NRF2, future studies may show it does. Many substances that may not show an increase in SOX9, may show an increase in osteoblast proliferation differentiation. If this is the case, than the substance may exhibit similar effects to SOX9 up-regulation.

Synergy can be positive and negative. When a substance exhibits synergy, you need less because it exhibits stronger effects or it may make the substances more easily absorbed by the body, penetrates the skin deeper or fights bad bacteria. Synergy can also be bad as it may interact with some medications and other herbs (*antagonism*). Trust your intuition and research for yourself if you are uncertain. Where you see this symbol (**), it denotes major synergism or increased effects.

Chart Legend	
S9 - increases SOX9 Gene	**SR - increases SIRT1**
NR - increases NRF2	**SY - exhibits synergy with....**
SU - available as a supplement	

Master table of synergistic and longevity gene upregulation substances and their synergy					
	S9	SR	NR	SU	SY
Herbs & Spices					
black cumin seed		*	*	*	
skullcap / baicalin	*	*	*	*	antibiotics / β-carotene /
chamomile flowers			*	*	apigenin-7-o-glucoside / ferruilic acid
coriander			*		chicory / probiotic / cumin seed oil / cefoperazone / gentamicin / tetracycline
rosemary		*	*	*	retinoic acid / antibiotics / ferulic & caffeic acids
cardamom			*	*	
astragalus			*	*	olive leaf / bitter melon / mulberry / phlorizin / atractylenolide / paeonia lactiflora / probiotics lactobacillus and bacillus cereus
mexican oregano			*		caffeic acid / chlorogenic acid / inalool, menthol,

					cinnamaldehyde, eugenol, thyme, rosemary, gentamicin, levofloxacin, polymyxin, kanamycin / antimicrobials
	S9	**SR**	**NR**	**SU**	**SY**
milk thistle (silybum marianum)	*	*	*	*	curcumin / silymarin / chrysin / artichoke / borututu
caraway seeds				*	retinoic acid / peppermint
fo ti he shou wu / polygonum multiflorum		*	*	*	reserveratrol / adriamycin
ginkgo biloba		*	*	*	green tea / aspirin / p. fruticosa leaf extract / fruticosa leaves
Plants					
green tea	*	*	*	*	lemon (**), imipenem / black pepper / lysine, arginine / quercetin, soy / theaflavin / cinnamon / ginger / acarbose / curcumin, /

					selenium / penicillin / sulindac / mint / honey / vitmin c / grape seed, gingko biloba, rosemary fruit of the persian oak (quercus brantii lindl.), amla (phyllanthus emblica l.), anar (punica granatum), dalchini (cinnamomum cassia, red onion,
	S9	**SR**	**NR**	**SU**	**SY**
longjing tea (*queen of green tea*)					Osmanthus fragrans flowers
rooibos tea		*			red palm oil
coffee / caffeine	*	*	*	*	isorhamnetin / catecholamines / ephedrine / sugar / aspirin / gallic acid / paclitaxel / isorhamnetin
dark chocolate / cacao		*	*	*	cinnamon / theobromine
coleus forskohlii / forskolin	*	*		*	phorbol ester / dexamethasone / carbachol
black tea		*	*	*	grape skins

lithospermum erythrorhizon / shikonin	*	*	*		antibiotics / erlotinib
comfrey			*	*	
	S9	**SR**	**NR**	**SU**	**SY**
lotus seed		*	*		phloridzin / green tea polyphenols
crabapple (*malus toringoides*)				*	
dendrobium officinale flower / dendrobium				*	
hawthorn berries		*	*	*	
burdock root	*	*	*	*	nettle leaves
lanceleaf tickseed (*coreopsis lanceolate flowers*) (*easy to grow at home anti-aging herb*)		*			
apple leaves			*	*	
coyol palm (*acrocomia aculeate*)				*	
beefsteak plant (*perilla frutescens*)			*	*	polygodial
wax gourd (benincasa hispida)		*	* *	*	

maitake mushroom			*	*	reishi & shiitake
maqui berry			*	*	citrus
	S9	**SR**	**NR**	**SU**	**SY**
Chrysanthe-mum		*	*	*	
erigeron breviscapus / breviscapine			*	*	
oroxylum indicum bark			*	*	acarbose
japanese knotweed			*		muscadine grape pomace extract
gotu kola (*centella asiatica*)		*	*	*	chloramphenicol / tetracycline / bacopa / vitamin e / curcumin
fructus psoraleae / psoralen	*	*	*		uva light / gold
dong quai (*angelica sinensis*)				*	
moringa oleifera leaves		**	**	*	moringa oleifera stem bark extract with ampicillin / amphotericin b
mango peel / mangiferin	*	*	**	*	antibiotics
citrus peel / nobiletin		*	*	*	atorvastatin / curcumin / sinensetin / sulforaphane **
passionfruit		*	*	*	
beet juice /		**	**	*	

betalain					
garlic / aged garlic		*	*		captopril / allicin / antibiotics / honey / fish
	S9	**SR**	**NR**	**SU**	**SY**
acai berry		***	*	*	olive leaf / reserveratrol
chickpeaas					beetroot / beet juice / betalain
Substances					
sulforaphane	*	*	**	**	selenium / mustard / luteolin / myricetin
luteolin		*	*	*	sulforaphane / quercetin / celecoxib
quercetin	*	*	*	*	luteolin / sulfamethoxazole / kaempferol / green tea / curcumin
curcumin	*	**	*	*	caryophyllene / lavocoxid / antibiotics / gemcitabine / xanthorrhizol / ginger / sulfinosine / epigallocatechin gallate / insulin / piperine / gallic acid / resveratrol

ellagic acid		*	*	*	lipoic acid / ferulic acid / quercetin / green tea / rutin
ferulic acid		*	*	*	ellagic acid / ascorbic acid / caspofungin / vitamin c isoferulic acid / coumaric acid / almond skins /
	S9	**SR**	**NR**	**SU**	**SY**
kaempferol	*	*	*	*	fluoroquinolones / 5-fluorouracil / quercetin / chrysin
genistein	*	**	**	*	tamoxifen / epigallocatechin gallate / curcumin / 5-fluorouracil/
horny goat weed / icariin	*	*	**	*	quercetin / rutin / formononetin / hispidulin
acacetin		*	*		
baicalein	**	*	*	*	daidzein / meropenem / trans-chalcone / gemcitabine / docetaxel / oxytetracycline / tetracycline / ceftazidime
vitamin c		*	*	*	vitamin e / green tea

vitamin e					vitamin c
carnosine	*	*	*	*	aminoguanidine , egcg, green tea, vitamin c, cod liver oil, grapeseed, bilberry
	S9	**SR**	**NR**	**SU**	**SY**
ginseng / ginsenoside rc	*	*	*	*	panax notoginseng saponins
resveratrol	*	*	*	*	cisplatin / oxyresveratrol / chrysin / curcumin / quercetin / chrysin / vitamin d / genistein / piperine / antiandrogen flutamide / polydatin / spermidine / fluorouracil / oxaliplatin / ginkgetin / cisplatin
hesperidin		*	*	*	mandarin peel extract / diazepam / chlorogenic acid
Epigallo-catechin	*	*	*	*	dl-cycloserine / digitonin / genistein (**) / luteolin /

					myricetin / daidzein / oxytetracycline / beta-lactams /tetracycline / cefotaxime
	S9	**SR**	**NR**	**SU**	**SY**
saffron		*	*	*	vanillin / pediculicide / low frequency electromagnetic fields
chondroitin sulfate	*	*	*	*	probiotics / vitamin c / glucosamine / curcumin / curcuma longa
linalool			*	*	chlorhexidine, cetylpyridinium, and triclosan (mouthwash only) / orange oil (external use only) / coumaric acid cumin seed
pinene			*		cineole/caryophy llene / pulegone, p-cymene, camphene (exterior only)
chitosan	***	*	*	*	thyme / whey protein / cinnamomum cassia presl /

					clove / zinc / caprylic acid, syzygium aromaticum (*preservation only*)
	S9	**SR**	**NR**	**SU**	**SY**
spermine	*	*	**	*	salicylic acid (**), putrescine
arginine		*	*	*	lactobacillus / phenylalanine /histidine/ ghrp-2 (kp 102) < hgh
lactobacillus acidophilus		*	*	*	probiotics / streptococcus thermophilus mk-10 / clove / cuminumcyminum l. oil / coriander seed / black rice / olive leaf extract
olive leaf extract		*	**	*	antibiotics / probiotics / metformin / hydroxytyrosol / oleuropein
vanillin / vanillic acid	***	*	**	*	catechins /clove
carnosic acid / rosemary	*	**	*		gentamicin / propolis (**) / curcumin / lutein / turmeric extract / tetracycline / lycopene (**), lutein and beta carotene

cinnamic acid		*	*		carvacrol (***) / vanillin / gallic acid / quercetin / thyme / polymyxin b
	S9	**SR**	**NR**	**SU**	**SY**
shikonin		*	*		antibiotics / erlotinib / metformin
bakuchiol		*	**	*	melatonin / garlic / vitmin c / ascorbyl tetraisopalmitate
limonene		*	*	*	berberine / carvone
mastic gum			*	*	pinene
anthocyanins		*	*		whey protein
cyanidin-3-glucoside		**	*	*	atorvastatin / acarbose / apple extracts / casein
gallic acid	*	*	*		caffeic acid (**) / protocatechuic acid / famotidine
apigenin-7-o-glucoside	**	*		*	luteolin / quercetin / cowpea extract / chrysanthemum / paclitaxel / pyrimethamine fluorouracil
peonidin	*				black rice germ
salicylic acid / aspirin					vorinostat / cyanidin / resveratrol / arginine /

					quercetin
Hydroxyl-cinnamic acids		*	*		leucine / sinapic acid
	S9	**SR**	**NR**	**SU**	**SY**
Hydroxyt-yrosol		*	*	*	
scutellarin	**	**	**		
silymarin / milk thistle	*	*	**	*	curcumin / chlorogenic acid / melatonin / chrysin / metformin
isorhamnetin			*		
Atractyle-nolide	**	*	**		
diosmin			*	*	fluorouracil
zeaxanthin			*	*	lutein / tocopherol / lycopene / zinc
beta-carotene					lycopene
phenylalanine				*	
uric acid				*	alpha-tocopherol
alpha-tocopherol					chlorogenic acid / beta-carotene / lycopene / gallic acid

Enhanced Bioavailability Of Nutrients In Foods		
food combination	**Effect**	**Synergisstic Substances**
banana & yoghurt and raw almonds **honey and garlic** **green leafy vegetables with lemon**	enhanced bioavailabilty of nutreints	vitamins a, e, c and zinc. vitamin c and iron
black pepper & green tea **green tea and lemon** **chocolate & raspberry**	immune system health	zinc, vitamin c, & vitamin d
fish with turmeric **turmeric with black pepper** **ginger and curcumin**	fights infection	zinc, vitamin c, & vitamin d
fish with garlic **olive oil with tomatoes**	strengthens body against chronic disease	vitamins a, k, d & zinc. calcium and selenium

grapes with onions **tomatoes with broccoli** **apples with berries** **meat with rosemary**		
chickpeas with beets	reproductive health	The B6 in the chickpeas increases the bioavailability of the magnesium in the beets. This can help ease symptoms of PMS and ADHD.
eating tomatoes, with boiled eggs carrots, and green leafy vegetables	eye health	increases absorption of carotenoids 3 to 9 times
extra virgin olive oil and apple-enriched dark chocolate	heart health	
black currant with rowanberry	controls blood glucose	

thyme honey with olive oil	significantly improves blood glucose	
red palm oil & rooibos tea	anti-inflammatory	
probiotics with prebiotics	enhances overall probiotics	
fish oil with tocopherols and rosemary extract	stronger antioxidants and also acts as a natural perservative	
almonds with dark chocolate	significant reductions in small dense low-density lipoproteins, which are lipoproteins that cause coronary heart disease	
carrots with hummus	The olive oil in the hummus increases the bioavailability of Vitamin A.	
tomatoes and olive oil	The olive oil enhances the bioavailabiilty of the beta-carotene up to five times more.	

Plant-based foods Nutrient Sources		
Food	**Sources**	**Generation Methods**
Protein	chickpeas, soy, soybeans	microbial fermentation (including mixed-cultured bacteria) which increases protein levels, Microalgae, products with essential amino acids.
Vitamin B12	breakfast cereals, yogurt and non-dairy milk alternatives.	fortification using natural vitamin B12-creating microorganisms via lactic fermentation, lupin fermentation used to create lupin tempeh, hydroponic cultivation methods, where crops grow in water that is enriched with vitamin B12.
Vitamin D	plant-based drinks, breakfast cereals, mushrooms, milk and eggs and Lichens D3 supplements.	vitamin D-biofortified eggs, UV irradiation of baker's yeast and mushrooms and lichens.
Iron	salt, dairy products, cereal-based products and milk.	biofortification, ferritin content enrichment, reduction of phytic acids, (*eg. soaking legumes or adding phytases during baking*), microencapsulation of an iron fortificant before mixing with foods, addition of ascorbic acid (eg. absorbic acid with Tofu).
Omega-3's	chia seeds, leafy green vegetables, flaxseeds, wheat germ, hempseeds and walnuts.	biofortificated foods such as eggs and dairy and eggs by incorporating algal or fish oil to cows' and hens' feed, cultured microalgae.
Calcium	breakfast cereals, plant-based drinks, Greek yogurt and kefir.	fermentation techniques involving mixed cultures of bacteria, Spirulina (Arthrospira sp).
Foods for Plant-Based Diets: Challenges and Innovations. Alexandra Alcorta et al. Feb 2021.		

Alchemical Effect and Season

The following information helps one to better understand the organic process of spiritualistic cycles and its corresponding alchemical influence. This allows one to more easily understand their stages of inner spiritual development. It can also be used as a guide to examine the influences of the 12 major constellations and their effects on heat such as cooking, frying or baking. Note: Due to axial precession, a new constellation starts rising behind the sun every 1,500 years. Please see my book titled: Farmer's Almanac Secrets. Correspondences of Feng Shui, Lunar and Alchemy for more information on axial precession.

Stage 1

At sunrise, the sun rises with the constellation ARIES behind it – March 20th to April 18th

Alchemical Theme: Calcination and Purification.

Element:: Fire

This marks the beginning of purification of one's consciousness via exposure to the fire element or by applying intense heat. It is a process of heating the body / aura to high temperatures, allowing for a loss of moisture, oxidation / reduction. This causes the breaking down of the whole into simple substances which prepare it to remove blockages and debris. It is the process of kundalini fire purification of the blood, flesh and bone. Kundalini energy begins in the tailbone, traveling upwards in the central vertical channel and chakras. It is a process that reoccurs many times towards spiritual enlightenment.

Stage 2

At sunrise, the sun rises with the constellation TAURUS behind it – April 19th to May 20th

Alchemical Theme: Transformation and Congelation.

Element: Earth

Congelation allows substances to congeal, or to thicken themselves, forming a new blueprint. Increases in viscosity reassembles the parts into a different energetic pattern via transformation. It occurs via a reduction in (applied) heat or by a change in bodily chemical reactions due to changes in temperature. The increase in viscosity may be powerful enough in that it solidifies or crystallizes new elements or substances. Congelation allows bodily transformation to take place via the spiritual integration of reassembled parts / patterns due to newly created layers of the body consciousness. Activities performed during this time will be more permanent and may see their value increase.

Stage 3

At sunrise, the sun rises with the constellation GEMINI behind it- May 21st to June 20th

Alchemical Theme: Synthesis and Fixation.

Element: Air

Fixation is where a previously volatile substance has been transformed into form / solid which is not affected by heat or fire. It separates substances or objects, rearranging them back to a new or original shape, or transforming it. In the body's energetic blueprint, it transforms it at the subatomic level. It is a continual cycle of polarity synthesis of unstable forces that become transformed into a more stable form at a higher level. Fixation is but one process required for the transformation of base substances. It is polarity integration between newly reassembled parts of subtracting or adding required patterns that become synthesized into the body consciousness.

Stage 4

At sunrise, the sun rises with the constellation CANCER behind it- June 21st to July 22nd

Alchemical Theme: Dismantling and Dissolution.

Element: Water

This is a dismantling phase from unnecessary energies, substances or objects which have been shifted from previous alchemical synthesis by the polarity integration of forces. It is where emotional body issues are resolved or dissolved into elements or parts, or a shifting in the field of energetic consciousness takes place. It is a deeply emotional phase, deeply buried in the subconscious. It brings to the surface the purging of deep soul wounds. In the phase of spiritual ascension. It may occur as a breaking up of the following - bonds, relationships, assembly or organization. Cancer exhibits very strong ancestral genetic pattern components which greatly influence emotional miasma in the body. Hence, the honoring of our ancestors during the sun or moon in Cancer is extremely beneficial. It represents dismantlement of unneeded or unnecessary primal forces and to resolve emotional blocks in the shifting terrain of energetic consciousness (emotional purging). A sign associated with feelings. It is a good time to plant crops or purchase stocks, commodities, collections or other long term investments. A good time to build friendships for growth.

Stage 5

At sunrise, the sun rises with the constellation LEO behind it-July 23rd to August 22nd

Alchemical Theme: Conversion and Digestion.

Element: Fire

Digestion is the process where low, but gentle heat is applied to the newly combined elements over a period of several weeks in order to completely alter its composition. This stage represents assimilation and absorption of the new substances. The energetic elements have now become shifted. Old unnecessary energies and emotional states have been eliminated by the action of enzymes and the newly converted energies move deeper into assimilation within the body. It is a time of digesting changes in one's consciousness which are the result of the completion of old energetic patterns. Allowing one to digest more of what they see in the world, one can resolve emotional pain allowing one to see the truth ever more clearly, which in the end allows one to experience greater acceptance of all things.

The approaching September 21st solstice is a good time to let things go and allow the spirit to experience fresh motivation. Think of it as the second spring or late spring of the year. It is an excellent time for public exhibitions, such as ceremonies, fundraisers, presentations, shows and entertainment.

Stage 6

At sunrise, the sun rises with the constellation VIRGO behind it– August 23rd to September 22nd

Alchemical Theme: Purity and Distillation

Element: Earth

This stage represents the internal energies and their continual process of purification which eventually separates the unnecessary parts, removing all contaminants. The end goal is to be able to achieve full concentration of the essential substance, which is where pure energetic consciousness comes from. It is the process of obtaining and accessing pure essence (via the inner spirit), and distilling the substance of spirit into the earth and body. It represents refinement of spirit and internal energies in relationship to the body.

Stage 7

At sunrise, the sun rises with the constellation LIBRA behind it– September 23rd to October 22nd

Alchemical Theme: Transmutation and Sublimation.

Element: Air

Transformation takes place through an observation of events or the experiencing of emotions. It sublimates them into higher substances. Hence one experiences spiritual harmony. As one elevates their consciousness, they learn to transform negative emotions. Observer consciousness allows one to witness their emotions or external events without having to identify with that particular emotion or event. Hence one has learned to observe emotions, thoughts and energies in a neutral tone. One has learned to increase their energetic harmony both inside and outside. This is greatly accelerated when one has learned to distill their inner spirit within.

Stage 8

At sunrise, the sun rises with the constellation SCORPIO behind it– October 23rd to November 21st

Alchemical Theme: Stillness and Separation.

Element: Water

A time of strong protection from interference / contaminates. This occurs via separation in order to attain purity. It is a time to experience the sequestering away from certain forces or influences. Isolation or wanting to sequester is a natural process, as in acquiring deep stillness. When it is acquired, one finds they have become one with God / Spirit. This allows one to sense the growth of their inner spirit. Isolation and its associative protective properties allows one to communicate with and germinate their spiritual consciousness towards fullness. An intuitive period, conductive to conducting research or ending a connection.

Stage 9

At sunrise, the sun rises with the constellation OPHIUCHUS (The Galactic Center / Center of the Milky Way) behind it- November 22nd to December 21st

Alchemical Theme: Wound Healing and Unification.

Element: Water/Aether

This phase represents the core of the Galactic Center. It is a place where spiritual waters pour themselves forth into the earth; bathing humanity. This powerful region of our galaxy represents a return to one's original geometry, the union of all polarity represented as a perfect blueprint and one of future energetic balance for humanity. Wisdom and knowledge is handed back to the truth seeker. The constellation in its highest expression represents the blueprint of divine perfection in a Human Being. It amplifies spiritual healing and is related to the symbol Phi and reemerging Sophia.

Stage 10

At sunrise, the sun rises with the constellation SAGITTARIUS behind it- December 22nd to February 18th

Alchemical Theme: Resurrection and Incineration.

Element: Fire

This phase represents the ascension process which occurs via incineration of the old and burning of the old decrepit energies, through the stages of a healing crisis. It is similar to a spiritual bonfire taking place in the lower self. Once the Cosmic aetheric energies and the divine blueprint have been connected to the body, massive healing begins. This reduces the old negative form and its related debris into ashes (the Phoenix that rises from the ashes). It is now reborn in a new form, which was previously incinerated in the bonfire. It is where spiritual alchemy undergoes phase transition, changing matter substance from blackened, to gray, and then finally to white. At times one senses death energy in the blackened state. It now enters the first stage of resurrection and great renewal. Good for higher learning, law, publishing and learning. Also a good time for adventure and sport.

Stage 11

At sunrise, the sun rises with the constellation CAPRICORN behind it–February 19th to March 19th

Alchemical Theme: Illumination and Fermentation.

Element: Earth

This represents the stage of fermentation which causes a shift in the production and the source of energy that is necessary for bodily functions. It is where an increased size and quality of the auric light body takes place. The body has become physically much stronger and more resistant to disease due to dryness, as in the example that earth's tropic of Capricorn is situated at a hot, dry latitude on our planet.

Fermentation is the process where different levels of enzymatic substances are created which the body uses to support a variety of beneficial shifting chemical and hormonal reactions. During the process of spiritual ascension, it may be associated with a strong physical shift in the immune, digestive and brain systems. It is where beneficial changes take place in how ATP production is made in the cells' mitochondria and how that energy becomes re-distributed to the body's cellular tissues. During this phase, new requirements of vital energies are taking place which increase the cell's ATP pump, causing increases in energy which help support the process of life. The increase of ATP allows a person to hold more light in their physical cells, which in turn increases their consciousness. A time that the immune system of the body is well-balanced.

Stage 12

At sunrise, the sun rises with the constellation AQUARIUS behind it– February 16 to March 11

Alchemical Theme: Virtues and Multiplication.

Element: Air

Multiplication greatly increases the concentrated refinement, energetic effectiveness and sphere of influence of the embodied biological spiritual light source. The blood is purified of karmic record and spiritualized via chemistry creating spiritual virtues. It occurs near the end of the magnum opus of our spiritual ascension alchemy. This is the symbolic state of which Christ is able to take any base substance and multiply it into many fish and bread loaves to bless the hungry. Through the access of pure embodied spiritual essence, the abundance of the spiritual light flows in and blesses the environment from the beacon of a Krystal body. It multiplies abundance and blessings for those who come into contact with it, as well as the being who embodies its source. It is about serving the group consciousness at its highest expression of spiritual ascension for all human beings. In traditional alchemy this is the turning of base metal into gold and silver, similar to what happens to the light body.

Stage 13

At sunrise, the sun rises with the constellation PISCES behind it– March 12 - April 18

Alchemical Theme: Perfection, Ascension and Christos-Sophia

Element: Water, Aether

This stage is attainment of complete preservation of the body via inner spirit, nourished in sound and light waves. It is the stage where one who has obtained the knowledge to spiritualizing their body and is able to transmute the corruptible physical into the incorruptible spirit. One understands that the universe is a holographic projection, and this realization allows one to break free of the negative karmic cycle in order to attain true freedom. One attains ultimate freedom from the prison planet mentality. This Great Work is the ultimate goal, the inner Christos-Sophia has been reached which makes perfect the complete divine human being. A good time for publishing, writing, prayer, QI Gong type practices, meditation or reflection. Artists and musicians are ruled by Pisces. If they avoid reality, the dark side of Pisces includes drugs, counterfeiting and related behavior.

Constellation and Subatomic Reaction	
Constellation	**Alchemical Effect**
Aries	Calcination
Taurus	Congelation
Gemini	Fixation
Cancer	Solution / Dissolution
Leo	Digestion
Virgo	Distillation
Libra	Sublimation
Scorpio	Separation
Sagittarius	Creation
Capricorn	Fermentation
Aquarius	Multiplication
Pisces	Projection

Recommended anti-aging reading list

- No limit to maximal lifespan in humans: how to beat a 122-year-old record. Mikhail V. Blagosklonny. Dec 2021.
- Antiaging agents: safe interventions to slow aging and healthy life span extension. Ji-Kai Liu.
- Botanical Compounds: Effects on Major Eye Diseases. Tuan-Phat Huynh et al. June 2013.
- State of the Art of Anthocyanins: Antioxidant Activity, Sources, Bioavailability, and Therapeutic Effect in Human Health. Noelia Tena et al. May 2020.
- The synergistic potential of various teas, herbs and therapeutic drugs in health improvement: A review. Florence Malongane et al. June 2017.
- Synergistic Effects of Natural Product Combinations in Protecting the Endothelium Against Cardiovascular Risk Factors. Muhammad Yousaf et al. July 2022.
- Nutraceutical potentials of synergic foods: a systematic review. Tharani Devi Natarajan et al. Dec 2019.
- Antiaging agents: safe interventions to slow aging and healthy life span extension. Ji-Kai Liu. May 2022.
- SIRT1 Activation by Natural Phytochemicals: An Overview. Concetta Iside et al. Aug 2020.
- Natural Nrf2 Activators from Juices, Wines, Coffee. Mallique Qader et al. Dec 2020.
- Synergistic Effects of Chinese Herbal Medicine. Western Sydney University. X Zhou et al. July 2016.
- Nutraceutical potentials of synergic foods: a systematic review. Tharani Devi Natarajan et al. Dec 2019.
- The synergistic potential of various teas, herbs and therapeutic drugs in health improvement: A review. F. Malongane et al. June 2017.
- SynergyFinder is a web-application that is used for finding synergistic drug combinations
- Antiaging agents: safe interventions to slow aging and healthy life span extension. Ji-Kai Liu. May 2022.
- SIRT1 Activation by Natural Phytochemicals: An Overview. Concetta Iside et al. Aug 2020.
- Impact of Polyphenolic-Food on Longevity: An Elixir of Life. An Overview. Rosaria Meccariello and Stefania D'Angelo. Mar 2021.

Cited Book References

1) The role of solar and geomagnetic activity in endothelial activation and inflammation in the NAS cohort. Jessica E. Schiff, et al. Jul 2022.
2) Association of Circulating C-Reactive Protein and Interleukin-6 with Longevity into the 80s and 90s: The Rancho Bernardo Study. Christina L. Wassel et al. Oct 2010.
3) Gregg Braden - How to Unlock Certain Mechanisms to Bring Health & Longevity Into Your Life. Gregg Braden Official
4) Short-term meditation training influences brain energy metabolism: A pilot study on 31P MR spectroscopy. Elke R. Gizewski et al. Jan 2021.
5) Centenarian Statistics. Thomas Perls MD, MPH. January 2023.
6) Sleep Is Good for Your Heart. Brooke Aggarwal. Sept 2022.
7) Impact of Polyphenolic-Food on Longevity: An Elixir of Life. An Overview. Rosaria Meccariello and Stefania D'Angelo et al. Apr 2021.
8) Association of sleep duration with risk of all-cause mortality and poor quality of dying in oldest-old people: a community-based longitudinal study. Chengbei Hou et al. Sept 2020.
9) Medicinal Plants of the Family Lamiaceae in Pain Therapy: A Review. Cristina M. Uritu et al. May 2018.
10) Medicinal Plants of the Family Lamiaceae in Pain Therapy: A Review. Cristina M. Uritu et al. May 2018.
11) Treatment of Glaucoma with Natural Products and Their Mechanism of Action: An Update. Ru Hui Sim et al. Jan 2022.
12) Methuselah, a Bristlecone Pine is Thought to be the Oldest Living Organism on Earth. Robert Hudson Westover, Public Affairs Specialist, USDA Forest Service in Forestry. Apr 21, 2011.
13) Chronicling Climate Change. Julia Rothchild. Smithsonian Science Education Center.
14) A natural mineral supplement provides relief from knee osteoarthritis symptoms: a randomized controlled pilot trial. Joy L Frestedt et al. Feb 2008.
15) Overexpression of SOX9 alleviates the progression of human osteoarthritis in vitro and in vivo. Yuanming Ouyang et al. Aug 2019.
16) Properties of Resveratrol: In Vitro and In Vivo Studies about Metabolism, Bioavailability, and Biological Effects in Animal Models and Humans. J. Gambini et al. June 2015.
17) Properties of Resveratrol: In Vitro and In Vivo Studies about Metabolism, Bioavailability, and Biological Effects in Animal Models and Humans. J. Gambini et al. June 2015.
18) Properties of Resveratrol: In Vitro and In Vivo Studies about Metabolism, Bioavailability, and Biological Effects in Animal Models and Humans. J. Gambini et al. June 2015.
19) Defense traits in the long-lived Great Basin bristlecone pine and resistance to the native herbivore mountain pine beetle. Barbara J. Bentz et al. Sept 2016.
20) 3-Carene, a Phytoncide from Pine Tree Has a Sleep-enhancing Effect by Targeting the GABAA-benzodiazepine Receptors Junsung Woo et al. Oct 2019.

21) Enhanced Emission of Monoterpene 3-Carene in Pinus densiflora Infected by Pine Wood Nematode and Characterization of 3-Carene Synthase. by Hwan-Su Hwang et al. Apr 2021.
22) Integrated view of plant metabolic defense with particular focus on chewing herbivores. David Wari, et al. Dec 2021.
23) A Review of Terpenes from Marine-Derived Fungi: 2015–2019. Minghua Jiang et al. June 2020.
24) Can a Jellyfish Unlock the Secret of Immortality?. New York Times.
25) The smell of longevity: a combination of Volatile Organic Compounds (VOCs) can discriminate centenarians and their offspring from age-matched subjects and young controls. Maria Conte et al. Dec 2019.
26) Production of Indole and Indole-Related Compounds by the Intestinal Microbiota and Consequences for the Host: The Good, the Bad, and the Ugly. by Naouel Tennoune et al. Apr 2022.
27) Piper nigrum Oil - Determination of Selected Terpenes for Quality Evaluation. Mei Wang et al. Feb 2019.
28) Investigation of Active Anti-Inflammatory Constituents of Essential Oil from Pinus koraiensis (Sieb. et Zucc.) Wood in LPS-Stimulated RBL-2H3 Cells. by Jiyoon Yang et al. May 2021.
29) Essential oils as anticancer agents: Potential role in malignancies, drug delivery mechanisms, and immune system enhancement. Mansi Sharma et al. Feb 2022.
30) Antioxidant and anti-aging potential of Juniper berry (Juniperus communis L.) essential oil in Caenorhabditis elegans model system. Swapnil Pandey et al. Sept 2018.
31) Analysis of the glutathione S-transferase (GST) gene family. Daniel W Nebert 1, Vasilis Vasiliou et al. 2004.
32) Essential oils as anticancer agents: Potential role in malignancies, drug delivery mechanisms, and immune system enhancement. Mansi Sharma et al. Feb 2022.
33) The development of theta and alpha neural oscillations from ages 3 to 24 years. Dillan Cellier et al. May 2021.
34) Normal and pathological changes in alpha rhythms. D Samson-Dollfus et al. Jun 1997.
35) The electroencephalogram in the middle-aged and the elderly. J Roubicek. April 1977.
36) Normal and pathological changes in alpha rhythms. D Samson-Dollfus et al. Jun 1997.
37) Theta and Alpha brainwaves are related to attention (Foxe and Snyder, 2011; Fries et al., 2001; Klimesch, 1999) and memory (Roux and Uhlhaas, 2014; Sauseng et al., 2009, 2005).
38) Effect of inhalation of isomers, (+)-α-pinene and (+)-β-pinene on human electroencephalographic activity according to gender difference. Minju Kim et al. Jan 2018.
39) Study on the Effect of Mentha × piperita L. Essential Oil on Electroencephalography upon Stimulation with Different Visual Effects. Shifan Lin et al. Jun 2022.
40) α-Pinene odor exposure enhances heat stress tolerance through Daf-16 in Caenorhabditis elegans. Naoko Ensaka and Kazuichi Sakamoto. Aug 2020.

41) Great Basin Bristlecone Pine Volatiles as a Climate Change Signal Across Environmental Gradients. Curtis A. Gray et al. Apr 2019.
42) Comparative evaluation of flavor compounds in fermented green and roasted coffee beans by solid phase microextraction-gas chromatography/mass spectrometry. Su-Jin Kim et al. Jul 2019.
43) Seasonality, Composition, and Antioxidant Capacity of Limonene/δ-3-Carene/(E)-Caryophyllene Schinus terebinthifolia Essential Oil Chemotype from the Brazilian Amazon: A Chemometric Approach. Bruna de Araújo Guimarães et al. June 2023.
44) Phenolic Compounds Present Schinus terebinthifolius Raddi Influence the Lowering of Blood Pressure in Rats. Lorena de Lima Glória et al. Oct 2017.
45) Metabolomic Analysis of Phytochemical Compounds from Ethanolic Extract of Lime (Citrus aurantifolia) Peel and Its Anti-Cancer Effects against Human Hepatocellular Carcinoma Cells. Pakkapong Phucharoenrak et al. Apr 2023.
46) Safety and efficacy of a feed additive consisting of an essential oil from Cinnamomum camphora (L.) J. Presl (camphor white oil) for use in all animal species. (FEFANA asbl). Nov 2021.
47) Dolomite - Uses, Side Effects, and More. Webmd. www.webmd.com/vitamins/ai/ingredientmono-29/dolomite
48) A natural mineral supplement provides relief from knee osteoarthritis symptoms: a randomized controlled pilot trial. Joy L Frestedt et al. Feb 2008.
49) Low concentration of 3-carene stimulates the differentiation of mouse osteoblastic MC3T3-E1 subclone 4 cells. Jong-Geun Jeong et al. Aug 2007.
50) The Antioxidant Content of Coffee and Its In Vitro Activity as an Effect of Its Production Method and Roasting and Brewing Time. Maciej Górecki and Ewelina Hallmann. Apr 2020.
51) Quantification of Caffeine and Chlorogenic Acid in Green and Roasted Coffee Samples Using HPLC-DAD and Evaluation of the Effect of Degree of Roasting on Their Levels. Shady Awwad, et al. Dec 2021.
52) Li H, Roxo M, Cheng X, Zhang S, Cheng H, Wink M. Pro-oxidant and lifespan extension effects of caffeine and related methylxanthines in Caenorhabditis elegans. Food Chem X. 2019;1:100005. doi: 10.1016/j.fochx.2019.100005.
53) Lublin A, Isoda F, Patel H, Yen K, Nguyen L, Hajje D, Schwartz M, Mobbs C. FDA-approved drugs that protect mammalian neurons from glucose toxicity slow aging dependent on cbp and protect against proteotoxicity. PLoS ONE. 2011;6(11):e27762.
54) Bridi JC, Barros AG, Sampaio LR, Ferreira JC, Antunes Soares FA, Romano-Silva MA. Lifespan extension induced by caffeine in Caenorhabditis elegans is partially dependent on adenosine signaling. Front Aging Neurosci. 2015;7:220. doi: 10.3389/fnagi.2015.00220.
55) Effect of food processing on antioxidants, their bioavailability and potential relevance to human health. Gamze Toydemir et al. June 2022.
56) Quantification of Caffeine and Chlorogenic Acid in Green and Roasted Coffee Samples Using HPLC-DAD and Evaluation of the Effect of Degree of Roasting on Their Levels. Shady Awwad, et al. Dec 2021.
57) Lonicera japonica extends lifespan and healthspan in Caenorhabditis elegans. Zhen-Zhou Yang et al. September 2018.

58) Temperatures of Meteoroids in Space. Butler, C. P.Meteoritics, volume 3, number 2, page 59.
59) Formation of amino acids on heating glycine with alumina. C P Ivanov and N N Slavcheva. April 1977.
60) Nonprotein Amino Acids from Spark Discharges and Their Comparison with the Murchison Meteorite Amino Acids. Yecheskel Wolman, et al. April 1972.
61) Nonprotein Amino Acids from Spark Discharges and Their Comparison with the Murchison Meteorite. Amino Acids. Yecheskel Wolman, et al. April 1972.
62) Panspermia. Wikipedia.
63) Life on Earth May Have Been Seeded by Comets. Rose Eveleth. March 8, 2013. Smithsonian Magazine.
64) On The Formation Of Dipeptides In Interstellar Model Ices. R. I. Kaiser Et Al. Feb 2013.
65) Zhuangzi text (Translation by Mair, 1998, pp. 6-7)
66) Amino acids in dew - Origin and seasonal variation. Edwin Scheller. Apr 2001.
67) Assessment of proline function in higher plants under extreme temperatures. A. Raza et al. Feb 2023.
68) Whitney, E.N. & Rolfes, S.R. (2015). Understanding Nutrition, 14th ed., Wadsworth, Cengage Learning, Belmont, CA.
69) Bernstein, M., & Munoz, N. (2016). Nutrition for the Older Adult, 2nd ed., Jones and Bartlett Publishers, Sudbury, MA.
70) Nutritional and lifestyle management of the aging journey: A narrative review. Amira Kassis et al. Jan 2023.
71) Factors in Aging that Effect the Bioavailability of Nutrients. Robert M. Russell. Apr 2001.
72) Nutrient absorption and intestinal adaptation with ageing. Trudy Woudstra and Alan B R Thomson. Feb 2002.
73) Providing Healthy and Safe Foods As We Age: Workshop Summary. Institute of Medicine (US) Food Forum. Washington (DC): National Academies Press (US); 2010.
74) Factors in Aging that Effect the Bioavailability of Nutrients. Robert M. Russell. Apr 2001.
75) Similarities and differences in the nutritional composition of nuts and seeds in Serbia. Margarita Dodevska, et al. Sept 2022.
76) Impact of Polyphenolic-Food on Longevity: An Elixir of Life. An Overview. Rosaria Meccariello and Stefania D'Angelo. Rosaria Meccariello and Stefania D'Angelo. April 2021.
77) A Greek island's secrets to long life, in 11 bullet points. Max Fisher. October 2012. Washington Post.
78) Youtube.com. - - - Farsight Intelligence Briefing: Mystery of My Age and Health - -
79) Healthy Lifestyle Recommendations: Do the Beneficial Effects Originate from NAD+ Amount at the Cellular Level?
80) Impact of Polyphenolic-Food on Longevity: An Elixir of Life. An Overview. Rosaria Meccariello and Stefania D'Angelo et al. Apr 2021.
81) Mechanisms of amino acid-mediated lifespan extension in Caenorhabditis

elegans. Clare Edwards et al. 2015.
82) Antioxidant and chelating capacity of Maillard reaction products in amino acid-sugar model systems: applications for food processing. Blanca A Mondaca-Navarro et al. Aug 2017.
83) Glycine supplementation extends lifespan of male and female mice. Richard A. Miller, et al. June 2019.
84) Dietary methionine restriction enhances metabolic flexibility and increases uncoupled respiration in both fed and fasted states. Barbara E. Hasek et al. Sept 2010.
85) Independent and Additive Effects of Glutamic Acid and Methionine on Yeast Longevity. Ziyun Wu et al. Nov 2013.
86) Glycine supplementation extends lifespan of male and female mice. Richard A. Miller, et al. June 2019.
87) Glycine supplementation extends lifespan of male and female mice. Richard A. Miller, et al. June 2019.
88) Plant Fortification of the Diet for Anti-Ageing Effects: A Review. Daljeet Singh Dhanjal et al. Sept 2020.
89) Health effects associated with consumption of unprocessed red meat: a Burden of Proof study. Haley Lescinsky et al. October 2022.
90) The impact of dietary protein intake on longevity and metabolic health. Munehiro Kitada et al May 2019.
91) Willcox BJ, Willcox DC, Todoriki H, Fujiyoshi A, Yano K, He Q, Curb JD, Suzuki M (October 2007). "Caloric restriction, the traditional Okinawan diet, and healthy aging: the diet of the world's longest-lived people and its potential impact on morbidity and life span". Annals of the New York Academy of Sciences. 1114: 434–55. doi:10.1196/annals.1396.037. PMID 17986602.
92) Food Products as Sources of Protein and Amino Acids—The Case of Poland. Hanna Górska-Warsewicz et al. Dec 2018.
93) Determination of the alkylpyrazine composition of coffee using stable isotope dilution-gas chromatography mass spectrometry (SIDA-GC-MS). Stephanie Pickard et al. Jul 2013.
94) Food Matrix Effects on Bioaccessibility ofβ-Carotene Can beMeasured in an in Vitro Gastrointestinal Model Carolien A. Van Loo-Bouwman, et al. The Journal of Agricultural and Food Chemistry.
95) Effect of Amino Acids on the Generation of Ginsenoside Rg3 Epimers by Heat Processing and the Anticancer Activities of Epimers in A2780 Human Ovarian Cancer Cells. Jun Yeon Park et al. Mar 2016.
96) D'Antona G, Ragni M, Cardile A, et al. Branched-chain amino acid supplementation promotes survival and supports cardiac and skeletal muscle mitochondrial biogenesis in middle-aged mice. Cell Metab. 2010;12(4):362–372. doi: 10.1016/j.cmet.2010.08.016
97) Branched-Chain Amino Acid Metabolism. Annual Review of Nutrition. Vol. 4:409-454 (Volume publication date July 1984).
98) The coming of age for branched-chain amino acids. Chen Gao, et al. Sept 2021.
99) The coming of age for branched-chain amino acids. Chen Gao, et al. Sept 2021.
100) Where to Find Leucine in Food and How to Feed Elderly With

Sarcopenia in Order to Counteract Loss of Muscle Mass: Practical Advice. Mariangela Rondanelli et al. Jan 2021.
101) The adverse metabolic effects of branched-chain amino acids are mediated by isoleucine and valine. Deyang Yu et al. May 2022.
102) Evaluation of a novel food composition database that includes glutamine and other amino acids derived from gene sequencing data. CM Lenders, et al. Jan 2009.
103) Protein content and amino acid composition of commercially available plant-based protein. isolatesStefan H. M. Gorissen et al. Aug 2018.
104) Influence of the protein digestion rate on protein turnover in young and elderly subjects. Martial Dangin et al. Branched-Chain Amino Acids Have Equivalent Effects to Other Essential Amino Acids on Lifespan and Aging-Related Traits in Drosophila. Paula Juricic, et al. Jan 2020.
105) Influence of the protein digestion rate on protein turnover in young and elderly subjects. Martial Dangin et al. Oct 2022.
106) Glycine Metabolism and Its Alterations in Obesity and Metabolic Diseases. Anaïs Alves et al. Jun 2019.
107) Glycine Metabolism and Its Alterations in Obesity and Metabolic Diseases. Anaïs Alves et al. Jun 2019.
108) Glycine supplementation extends lifespan of male and female mice. Richard A. Miller, et al. June 2019.
109) Branched-Chain Amino Acids Have Equivalent Effects to Other Essential Amino Acids on Lifespan and Aging-Related Traits in Drosophila. Paula Juricic, et al. Jan 2020.
110) Lifelong restriction of dietary branched-chain amino acids has sex-specific benefits for frailty and lifespan in mice. Nicole E. Richardson et al. July 2021.
111) Karusheva Y, Koessler T, Strassburger K, Markgraf D, Mastrototaro L, Jelenik T, et al. Short-term dietary reduction of branched-chain amino acids reduces meal-induced insulin secretion and modifies microbiome composition in type 2 diabetes: a randomized controlled crossover trial. Am J Clin Nutr 2019;110(5):1098–107.
112) Effect of dietary glycine on methionine metabolism in rats fed a high-methionine diet. K Sugiyama et al. June 1987.
113) Glycine and Longevity. https://novoslabs.com/
114) Dietary glycine supplementation mimics lifespan extension by dietary methionine restriction in Fisher 344 rats. Joel Brind et al. Apr 2011.
115) Glycine supplementation extends lifespan of male and female mice. Richard A. Miller et al. Mar 2019.
116) Glycine supplementation extends lifespan of male and female mice. Richard A. Miller et al. Mar 2019.
117) Multifarious Beneficial Effect of Nonessential Amino Acid, Glycine: A Review. Meerza Abdul Razak, et al. 2017.
118) Dietary glycine supplementation mimics lifespan extension by dietary methionine restriction in Fisher 344 rats. Joel Brind et al. Apr 2011.
119) Effect of arginine:lysine and glycine:methionine intake ratios on dyslipidemia and selected biomarkers implicated in cardiovascular disease: A study with hypercholesterolemic rats. Ravula Venkatesh et al. July 2017.
120) https://www.myfooddata.com

121) Food Products as Sources of Protein and Amino Acids—The Case of Poland. Hanna Górska-Warsewicz et al. Dec 2018.
122) Betaine in Cereal Grains and Grain-Based Products. Bojana Filipčev et al. Mar 2018.
123) Glycine Metabolism and Its Alterations in Obesity and Metabolic Diseases. Anaïs Alves et al. Jun 2019.
124) https://tools.myfooddata.com
125) A commentary on the paper: 'Evaluation of spice and herb as phytoderived selective modulators of human retinaldehyde dehydrogenases using a simple in vitro method'. Anna Bilska-Wilkosz. Jan 2022.
126) Dietary and herbal supplements for fatigue: A quality assessment of online consumer health information. Jeremy Y Ng, et al. May 2021.
127) Characterization, Variables, and Antioxidant Activity of the Maillard reaction in a Fructose–Histidine Model System. Pengli Liu et al. Dec 2018.
128) Insights into flavor and key influencing factors of Maillard reaction products: A recent update. Shuyun Liu, Hanju Sun et al. Sept 2022.
129) Insights into flavor and key influencing factors of Maillard reaction products: A recent update. Shuyun Liu, Hanju Sun et al. Sept 2022.
130) Lotus - the natural products occurrence database. - 20R_-Ginsenoside-Rg3.
131) Ginsenosides Rk1 and Rg5 inhibit transforming growth factor-β1-induced epithelial-mesenchymal transition and suppress migration, invasion, anoikis resistance, and development of stem-like features in lung cancer. Hyunhee Kim et al. Jan 2021.
132) Effect of Different Cooking Methods on Polyphenols, Carotenoids and Antioxidant Activities of Selected Edible Leaves. K. D. Prasanna P. Gunathilake et al. Sept 2018.
133) Li Ching-Yuen: The 256 year-old Herbalist lived on a diet of Herbs. https://www.newsgram.com/
134) Li Ching-Yuen: The 256 year-old Herbalist lived on a diet of Herbs. https://www.newsgram.com/
135) Time Magazine. CHINA: Tortoise-Pigeon-Dog. Monday, May 15, 1933.
136) Zaro Agha Dies At Mooted Age Of Ighears Scientists to Test Longevity Claim of Turk By Autopsy. San Pedro News Pilot, Volume 7, Number 100, 29 June 1934.
137) Lysine-glucose Maillard reaction products promote longevity and stress tolerance in Caenorhabditis elegans via the insulin/IGF-1 signaling pathway. Yokoyama et al. Dec 2021.
138) Demonstration of antioxidant and anti-inflammatory bioactivities from sugar-amino acid Maillard reaction products. David D Kitts et al. Jul 2012.
139) Impact of infrared and dry air roasting on the oxidative stability, fatty acid composition, Maillard reaction products and other chemical properties of black cumin (Nigella sativa L.) seed oil. Kanchan Suri et al. Oct 2019.
140) Coffee Extends Yeast Chronological Lifespan through Antioxidant.

Jadwiga Czachor et al. Dec 2020.
141) Aging - Oxidative stress, antioxidants and computational modeling. Umm-e-Ammara Warraich, et al. May 2020.
142) Li H, Roxo M, Cheng X, Zhang S, Cheng H, Wink M. Pro-oxidant and lifespan extension effects of caffeine and related methylxanthines in Caenorhabditis elegans. Food Chem X. 2019;1:100005. doi: 10.1016/j.fochx.2019.100005.
143) Lublin A, Isoda F, Patel H, Yen K, Nguyen L, Hajje D, Schwartz M, Mobbs C. FDA-approved drugs that protect mammalian neurons from glucose toxicity slow aging dependent on cbp and protect against proteotoxicity. PLoS ONE. 2011;6(11):e27762.
144) Bridi JC, Barros AG, Sampaio LR, Ferreira JC, Antunes Soares FA, Romano-Silva MA. Lifespan extension induced by caffeine in Caenorhabditis elegans is partially dependent on adenosine signaling. Front Aging Neurosci. 2015;7:220. doi: 10.3389/fnagi.2015.00220.
145) The Review of Anti-aging Mechanism of Polyphenols on Caenorhabditis elegans. Limin Liu1 et al. July 2021.
146) Determination of Phloridzin and Other Phenolic Compounds in Apple Tree Leaves, Bark, and Buds Using Liquid Chromatography with Multilayered Column Technology and Evaluation of the Total Antioxidant Activity. Anežka Adamcová et al. Feb 2022.
147) Dietary intake of phloridzin from natural occurrence in foods. Katherine E. Niederberger et al. Jan 2020. Cambridge University.
148) In vitro antioxidant and cytoprotective properties of Maillard reaction products from phloridzin-amino acid model systems. Linna Han et al. Jan 2018.
149) Radiolytic Cyclization Products of Phloridzin as Potent Anti-Glycation Agents. Gyeong Han Jeong and Tae Hoon Kim. July 2020.
150) Cytotoxic effects of phloridzin. A. Cansu Kilit and Esra Aydemir. Northwestern Med J.2022;2(1):16-22. doi: 10.54307/NWMJ.2022.28247. Jan 2022.
151) Sesamin extends lifespan through pathways related to dietary restriction in Caenorhabditis elegans. Yumiko Nakatani et al. Apr 2018.
152) Bioactive Phytochemicals with Anti-Aging and Lifespan Extending Potentials in Caenorhabditis elegans. Nkwachukwu Oziamara Okoro, et al. Dec 2021.
153) Sesame Lignans Suppress Age-Related Cognitive Decline in Senescence-Accelerated Mice. Satomi Shimoyoshi et al. Jul 2019.
154) Effects of roasting temperature and duration on color and flavor of a sesame oligosaccharide-protein complex in a Maillard reaction model. Qing Guo et al. Dec 2022.
155) Clinical outcomes of two patients with a novel pathogenic variant in ASNS: response to asparagine supplementation and review of the literature. Rosanne Sprute et al. May 2019.
156) Metabolomic Profiling of Fresh Goji (Lycium barbarum L.) Berries from Two Cultivars Grown in Central Italy: A Multi-Methodological Approach. Mattia Spano et al. Sept 2021.
157) Measurement of Fructose–Asparagine Concentrations in Human and Animal Foods. Jikang Wu et al. Dec 2017.

158) As Extracellular Glutamine Levels Decline, Asparagine Becomes an Essential Amino Acid. Natalya N. Pavlova et al.
159) Xian: Immortality in the Daoist Tradition. Pablo Vazquez. 2019.
160) Xian: Immortality in the Daoist Tradition. Pablo Vazquez. 2019.
161) Zhang Sanfeng. Wikipedia.
162) Understanding Modern Internal Alchemy Series: A (very) Brief History of Secular Internal Alchemy. Robert J. Coons. Aug 2022.
163) Pengzu. Wikipedia.
164) Zhang Guolao. Wikipedia.
165) Lü Dongbin. Wikipedia.
166) Xian: Immortality in the Daoist Tradition. Pablo Vazquez. 2019.
167) Xian: Immortality in the Daoist Tradition. Pablo Vazquez. 2019.
168) Han Xiang Zi. Wikipedia.
169) Differential Effects of Red Meat/Refined Grain Diet and Dairy/Chicken/Nuts/Whole Grain Diet on Glucose, Insulin and Triglyceride in a Randomized Crossover Study. Yoona Kim, et al. Oct 2016.
170) Reduced Circulating Insulin Enhances Insulin Sensitivity in Old Mice and Extends Lifespan. Nicole M. Templeman et al. July 2017.
171) Innovative Immunization Strategies for Antivenom Development. Erick Bermúdez-Méndez et al. Nov 2018.
172) Cheon JM, Kim DI, Kim KS. Insulin sensitivity improvement of fermented Korean Red Ginseng (Panax ginseng) mediated by insulin resistance hallmarks in old-aged ob/ob mice. J Ginseng Res. 2015; 39:331–37. 10.1016/j.jgr.2015.03.005.
173) Fermenting Red Ginseng Enhances Its Safety and Efficacy as a Novel Skin Care Anti-Aging Ingredient: In Vitro and Animal Study. Hyun-Sun Lee et al. Nov 2012.
174) Effect of Amino Acids on the Generation of Ginsenoside Rg3 Epimers by Heat Processing and the Anticancer Activities of Epimers in A2780 Human Ovarian Cancer Cells. Jun Yeon Park et al. Mar 2016.
175) D'Antona G, Ragni M, Cardile A, et al. Branched-chain amino acid supplementation promotes survival and supports cardiac and skeletal muscle mitochondrial biogenesis in middle-aged mice. Cell Metab. 2010;12(4):362–372. doi: 10.1016/j.cmet.2010.08.016.
176) Stereoisomers ginsenosides-20(S)-Rg3 and -20(R)-Rg3 differentially induce angiogenesis through peroxisome proliferator-activated receptor-gamma. Hoi-Hin Kwok et al. Apr 2012.
177) Adooq Biosciences. 20(R)Ginsenoside Rg3. Catalog No.: A14731. www.adooq.com/20-r-ginsenoside-rg3.html
178) Preparation of Ginsenoside Rg3 and Protection against H2O2-Induced Oxidative Stress in Human Neuroblastoma SK-N-SH Cells. Gang Li et al. May 2014.
179) Ginseng for an eye: effects of ginseng on ocular diseases. Jisu Kim et al. Sept 2018. Journal of Ginseng Research.
180) Aging - Oxidative stress, antioxidants and computational modeling. Umm-e-Ammara Warraich, et al. May 2020.
181) Cytoprotective Effect of 20(S)-Rg3 on Benzo[a]pyrene-Induced DNA Damage. Po Ying Poon et al. Jan 2012.
182) Lotus - the natural products occurrence database. www.

pubchem.ncbi.nlm.nih.gov/compound/20R_-Ginsenoside-Rg3#section=Structures.
183) Lotus - the natural products occurrence database. www. pubchem.ncbi.nlm.nih.gov/compound/Ginsenoside-Rg5.
184) A systematic exploration of ginsenoside Rg5 reveals anti-inflammatory functions in airway mucosa cells. Hyojin Heo et al. Jan 2023.
185) Ginsenoside Rh4 Suppressed Metastasis of Lung Adenocarcinoma via Inhibiting JAK2/STAT3 Signaling. Yan Zhang et al. Feb 2022.
186) An analysis of the combination frequencies of constituent medicinal herbs in prescriptions for the treatment of bone and joint disorder in Korean medicine: determination of a group of candidate prescriptions for universal use. Yoo Kyoung Han et al. Sept 2017.
187) Ginseng for an eye: effects of ginseng on ocular diseases. Jisu Kim et al. Aug 2018. Journal of Ginseng Research.
188) The Role of SOX Transcription Factors in Ageing and Age-Related Diseases. Milena Stevanovic et al. Jan 2023. Protective Effect of Ginsenoside Rb1 on Hydrogen Peroxide-induced Oxidative Stress in Rat Articular Chondrocytes.
189) Research Progress on the Antiosteoarthritic Mechanism of Action of Natural Products. Mingzhu Gao et al. Sept 2021.
190) Ginsenoside Rb1 inhibits monoiodoacetate-induced osteoarthritis in postmenopausal rats through prevention of cartilage degradation. Adithan Aravinthan et al. Jan 2020.
191) Ginsenoside Rb1 from Panax ginseng attenuates monoiodoacetate-induced osteoarthritis by inhibiting miR-21-5p/FGF18-mediated inflammation. Jingjie Luan et al. July 2022.
192) Ginsenoside Rg1 Prevents H2O2-induced Lens Opacity. Guowei Zhang et al. Aug 2021.
193) Ginseng for an eye: effects of ginseng on ocular diseases. Jisu Kim et al. Nov 2018.
194) Protective Effect of Ginsenoside Rb1 on Hydrogen Peroxide-induced Oxidative Stress in Rat Articular Chondrocytes.
195) Processed Panax ginseng, sun ginseng, inhibits the differentiation and proliferation of 3T3-L1 preadipocytes and fat accumulation in Caenorhabditis elegans. Author links open overlay panelHyojin Lee et al. July 2017.
196) Effect of Amino Acids on the Generation of Ginsenoside Rg3 Epimers by Heat Processing and the Anticancer Activities of Epimers in A2780 Human Ovarian Cancer Cells. Jun Yeon Park et al. Aug 2015.
197) Enhanced healing of full-thickness burn wounds using di-rhamnolipidTamara Stipcevic et al. Feb 2007.
198) Pathway elucidation of bioactive rhamnosylated ginsenosides in Panax ginseng and their de novo high-level production by engineered Saccharomyces cerevisiae. Chaojing Li et al. Aug 2022.
199) Pharmacology of ginsenosides: a literature review. Kar Wah Leung. Alice Sze-Tsai Wong et al. June 2010.
200) Two new acetylated ginsenosides from the roots of Panax quinquefolium. Ji Ming Jia et al. Sept 2008.
201) Cardioprotective potential of N,α-L-rhamnopyranosyl vincosamide,

an indole alkaloid, isolated from the leaves of Moringa oleifera in isoproterenol induced cardiotoxic rats: in vivo and in vitro studies. Sunanda Panda et al. Feb 2013.

202) Acacetin 7-O-α-l-rhamnopyranosyl (1–2) β-D-xylopyranoside Elicits Life-span Extension and Stress Resistance in Caenorhabditis elegans. Jyotsna Asthana, et al. Oct 2015.

203) Health Benefits of Edible Leave of Premna serratifolia L.. Kris Herawan Timotius et al. Sept 2021.

204) Evaluation of Antioxidative and Neuroprotective Activities of Total Flavonoids From Sea Buckthorn (Hippophae rhamnoides L. Zheng Wang et al. June 2022.

205) Healing effect of sea buckthorn, olive oil, and their mixture on full-thickness burn wounds. Mitra Edraki et al. July 2014.

206) Antioxidant, DNA damage protective, neuroprotective, and α-glucosidase inhibitory activities of a flavonoid glycoside from leaves of Garcinia gracilis. Chonlakan Supasuteekul, et al. Mar 2016.

207) New triterpenoids and other constituents from the fruits of Benincasa hispida (Thunb.) Cogn. Xiao-Na Han et al. Dec 2013.

208) Formulation and in vitro evaluation of the topical antiageing preparation of the fruit of Benincasa hispida. Vidya Sabale et al. Jul 2011.

209) 100'th update on fly longevity experiments. www.cryonet.org/cgi-bin/dsp.cgi?msg=22367

210) 101'st update on fly longevity experiments. www.cryonet.org/cgi-bin/dsp.cgi?msg=22414

211) Extract of Wax Gourd Peel Prevents High-Fat Diet-Induced Hyperlipidemia in C57BL/6 Mice via the Inhibition of the PPARγ Pathway. Ming Gu et al. Feb 2013.

212) Nitric oxide-dependent hypotensive effects of wax gourd juice. Miki Nakashima et al. Nov 2011.

213) Antioxidant and angiotension-converting enzyme inhibition capacities of various parts of Benincasa hispida (wax gourd). Hui-Yu Huang et al. June 2004.

214) Formulation and in vitro evaluation of the topical antiageing preparation of the fruit of Benincasa hispida. Vidya Sabale et al. July 2011.

215) Valorization of Polysaccharides From Benincasa hispida: Physicochemical, Moisturizing, and Antioxidant Skincare Properties. Qian Wang et al. June 2022.

216) Kundur [Benincasa hispida (Thunb.) Cogn.]: A potential source for valuable nutrients and functional foods. Nurul Aqilah Mohd Zaini et al. Aug 2011.

217) A Literature-Based Update on Benincasa hispida (Thunb.) Cogn.: Traditional Uses, Nutraceutical, and Phytopharmacological Profile. Muhammad Torequl Islams et al. Dec 2021.

218) Huang H. Y., Huang J. J., Tso T. K., Tsai Y. C., Chang C. K. Antioxidant and angiotension converting enzyme inhibition capacities of various parts of Benincasa hispida (wax gourd) Food/Nahrung . 2004;48(3):230–233. doi: 10.1002/food.200300428

219) Kundur [Benincasa hispida (Thunb.) Cogn.]: A potential source for valuable nutrients and functional foods. Nurul Aqilah Mohd Zaini et al.

Aug 2011.
220) Acacetin 7-O-α-l-rhamnopyranosyl (1–2) β-D-xylopyranoside Elicits Life-span Extension and Stress Resistance in Caenorhabditis elegans. Jyotsna Asthana, et al. Sept 2016.
221) Anti-aging Effects of Camellia Japonica flower extract on a pollutant-induced stress. Seoungwoo Shin et al. May 2018.
222) Camellianoside, a Novel Antioxidant Glycoside from the Leaves of Camellia japonica. Ken-ichi Onodera et al. Aug 2006.
223) Camellia japonica: A phytochemical perspective and current applications facing its industrial exploitation. Antia G. Pereira, et al. Mar 2022.
224) Lipid Characteristics of Camellia Seed Oil. Wei Zeng and Yasushi Endo. July 2019.
225) Olives and Olive Oil in Health and Disease Prevention. Victor R. Preedy and Ronald Ross Watson.
226) Camellia japonica: A phytochemical perspective and current applications facing its industrial exploitation. Antia G. Pereira, et al. Mar 2022.
227) Plants and phytochemicals for Huntington's disease. Sunayna Choudhary, Puneet Kumar,1 and Jai Malik et al. July 2013.
228) Extraction And Characterisation Of The Essential Oil From Centella Asiatica (Pennywort) Growing In South Africa. Melissa Claudia Florczak. A Dissertation Submitted To The Faculty Of Science. University Of The Witwatersrand, Johannesburg, In Fulfilment Of The Requirements For The Degree Of Master Of Science. Johannesburg, 2014.
229) Lotus - the natural products occurrence database. www.pubchem.ncbi.nlm.nih.gov/compound/20R_-Ginsenoside-Rg3#section=Structures
230) Lotus - the natural products occurrence database. www.pubchem.ncbi.nlm.nih.gov/compound/Ginsenoside-Rg5
231) Li Ching-Yuen: The 256 year-old Herbalist lived on a diet of Herbs. www.newsgram.com/
232) Time Magazine. CHINA: Tortoise-Pigeon-Dog. Monday, May 15, 1933.
233) Effect of Different Cooking Methods on Polyphenols, Carotenoids and Antioxidant Activities of Selected Edible Leave. K. D. Prasanna P. Gunathilake et al. Sept 2018.
234) Centella asiatica and Its Metabolite Asiatic Acid: Wound Healing Effects and Therapeutic Potential. Lúcio Ricardo Leite Diniz et al. Feb 2023.
235) Asiatic acid attenuates hypertrophic and fibrotic differentiation of articular chondrocytes via AMPK/PI3K/AKT signaling pathway. Na Liu et al. May 2020.
236) Potential Antiaging Effects of DLBS1649, a Centella asiatica Bioactive Extract. Agung H Karsono et al. Aug 2021.
237) COL2A1 gene. www.medlineplus.gov
238) Extracellular Matrix Proteins—Advances in Research and Application: 2012 Edition. Scholarly Editions, 26 Dec 2012.
239) The Role of SOX Transcription Factors in Ageing and Age-Related Diseases. Milena Stevanovic. Jan 2023.

240) Isorhapontigenin Modulates SOX9/TOLLIP Expression to Attenuate Cell Apoptosis and Oxidative Stress in Paraquat-Induced Acute Kidney Injury. Qiang Zheng. June 2022.

241) Extracellular Matrix Proteins—Advances in Research and Application: 2012 Edition. Scholarly Editions, 26 Dec 2012

242) Regulation and function of SOX9 during cartilage development and regeneration. Haengseok Song 1, Keun-Hong Park . May 2020.

243) SOX9 keeps growth plates and articular cartilage healthy by inhibiting chondrocyte dedifferentiation/osteoblastic redifferentiation. Abdul Haseeb. Feb 2021.

244) Baicalin promotes extracellular matrix synthesis in chondrocytes via the activation of hypoxia-inducible factor-1α. Pengzhen Wang et al. Dec 2020.

245) The genomes of medicinal skullcaps reveal the polyphyletic origins of clerodane diterpene biosynthesis in the family Lamiaceae. Haixiu Li et al. Mar 2023.

246) Micropropagation of Ajuga bracteosa, a medicinal herb. Shivanee Kaul et al. Apr 2013.

247) A Review of Diterpenes from Marine-Derived Fungi: 2009–2021. Peng Qiu, et al. Nov 2022.

248) Research Progress on the Antiosteoarthritic Mechanism of Action of Natural Products. Mingzhu Gao, et al. Sept 2021.

249) Effect of Active Ingredients of Chinese Herbal Medicine on the Rejuvenation of Healthy Aging: Focus on Stem Cells. Chen Wang et al. Jul 2020.

250) Isopsoralen Induces Differentiation of Prechondrogenic ATDC5 Cells via Activation of MAP Kinases and BMP-2 Signaling Pathways. Liang Li et al. May 2012.

251) UP256 Inhibits Hyperpigmentation by Tyrosinase Expression/Dendrite Formation via Rho-Dependent Signaling and by Primary Cilium Formation in Melanocytes. Min Cheol Kang et al. Jul 2020.

252) Lipid availability determines skeletal progenitor cell fate via SOX9. Nick van Gastel, et al. Feb 2020.

253) Mechanisms and disease implications of sirtuin-mediated autophagic regulation. In Hye Lee. Sept 2019.

254) SIRT1 in the brain—connections with aging-associated disorders and lifespan. Fanny Ng,. Mar 2015.

255) Healthy Lifestyle Recommendations: Do the Beneficial Effects Originate from NAD+ Amount at the Cellular Level?. Borut Poljsak et al. Dec 2020.

256) SIRT1 in the brain—connections with aging-associated disorders and lifespan. Fanny Ng et al. Mar 2015.

257) SIRT1 in the brain—connections with aging-associated disorders and lifespan. Fanny Ng et al. Mar 2015.

258) Impact of Nutrition on Short-Term Exercise-Induced Sirtuin Regulation: Vegans Differ from Omnivores and Lacto-Ovo Vegetarians. Arne Björn Potthast et al. Apr 2020.

259) Impact of Nutrition on Short-Term Exercise-Induced Sirtuin Regulation: Vegans Differ from Omnivores and Lacto-Ovo Vegetarians.

Arne Björn Potthast et al. Apr 2020.
260) Sirtuin-1 (SIRT1) is required for promoting chondrogenic differentiation of mesenchymal stem cells. Constanze Buhrmann et al. Aug 2014.
261) Fasting and fasting mimicking treatment activate SIRT1/LXRα and alleviate diabetes-induced systemic and microvascular dysfunction. Sandra S. Hammer et al. Mar 2021.
262) Fasting and fasting mimicking treatment activate SIRT1/LXRα and alleviate diabetes-induced systemic and microvascular dysfunction. Sandra S. Hammer et al. Mar 2021.
263) The Protective Effect of Baicalin Against Lead-Induced Renal Oxidative Damage in Mice. Zecai Zhang et al. Jan 2017.
264) Protective effect of quercetin on lead-induced oxidative stress and endoplasmic reticulum stress in rat liver via the IRE1/JNK and PI3K/Akt pathway. Chan-Min Liu et al. Dec 2012.
265) Philosophers' Stone. www.encyclopedia.com
266) Effects of Urin Cow Dosage on Growth and Production of Sorgum Plant (Sorghum Bicolor L) on Peat Land. Sri Utami Lestari and Andi Andrian. 2017.
267) Peptide profiling in cow urine reveals molecular signature of physiology-driven pathways and in-silico predicted bioactive properties. Rohit Kumar, et al. June 2021.
268) Amino acids in dew - Origin and seasonal variation. Edwin Scheller. Apr 2001.
269) Cow urine distillate as bioenhancer. Gurpreet Kaur Randhawa. Dec 2010.
270) Beneficial of Coriander Leaves (Coriandrum sativum L.) to Reduce Heavy Metals Contamination in Rod Shellfish. Winarti, S. et al. Jan 2018.
271) Cow urine distillate as bioenhancer. Gurpreet Kaur Randhawa. Dec 2010.
272) Cow urine distillate as bioenhancer. Gurpreet Kaur Randhawa. Dec 2010.
273) An assessment of university students and staff perceptions regarding the use of human urine as a valuable soil nutrient in South Africa. LL Mugivhisa and JO Olowoyo. Sept 2015.
274) An assessment of university students and staff perceptions regarding the use of human urine as a valuable soil nutrient in South Africa. LL Mugivhisa and JO Olowoyo. Sept 2015.
275) Biostimulant Effects of an Aqueous Extract of Duckweed (Lemna minor L.) on Physiological and Biochemical Traits in the Olive Tree. Luca Regni et al. Dec 2021.
276) The Prospect of Duckweed in Pig Nutrition: A Review. Francisca Rumosa Gwaze and Marizvikuru Mwale. Oct 2015.
277) Dietary Arginine Supplementation Enhances the Growth of Milk-Fed Young Pigs . Sung Woo Kim et al. Mar 2004.
278) Magnetic nanostructures functionalized with a derived lysine coating applied to simultaneously remove heavy metal pollutants from environmental systems. Olivija Plohl et al. Jan 2021.
279) Physiological Effects of Dietary Amino Acids on Gut Health and

Functions of Swine. Zhongyue Yang and Shengfa F. Liao et al. June 2019.
280) Natural nutraceuticals for enhancing yogurt properties: a review. Ahmed K. Rashwan et al. March 2023.
281) Moderate Red Wine Consumption Increases the Expression of Longevity-Associated Genes in Controlled Human Populations and Extends Lifespan in Drosophila melanogaster. Juan Gambini et al. Feb 2021.
282) Resveratrol: A Natural Compound with Pharmacological Potential in Neurodegenerative Diseases. Héctor I. Rocha-González, et al. Oct 2008.
283) Resveratrol: A Natural Compound with Pharmacological Potential in Neurodegenerative Diseases. Héctor I. Rocha-González, et al. Oct 2008.
284) Resveratrol: A Double-Edged Sword in Health Benefits. Bahare Salehi et al. Sept 2018.
285) Resveratrol: A Natural Compound with Pharmacological Potential in Neurodegenerative Diseases. Héctor I. Rocha-González, et al. Oct 2008.
286) Resveratrol: A Natural Compound with Pharmacological Potential in Neurodegenerative Diseases. Héctor I. Rocha-González, et al. Oct 2008.
287) Studies on the metabolism of amino acids and related compounds in vivo. I. Toxicity of essential amino acids, individually and in mixtures, and the protective effect of l-arginine. Piero Gullino et al. Oct 1956.
288) The Effects of Consuming Amino Acids L-Arginine, L-Citrulline (and Their Combination) as a Beverage or Powder, on Athletic and Physical Performance: A Systematic Review. Siphamandla Nyawose et al. Aug 2022.
289) L-arginine supplementation protects exercise performance and structural integrity of muscle fibers after a single bout of eccentric exercise in rats. Yulia N Lomonosova et al. Apr 2014.
290) L-Arginine Exerts Excellent Anti-Stress Effects on Stress-Induced Shortened Lifespan, Cognitive Decline and Depression. Monira Pervin et al. Jan 2021.
291) Formation of amino acids on heating glycine with alumina. C P Ivanov and N N Slavcheva. April 1977.
292) Independent and Additive Effects of Glutamic Acid and Methionine on Yeast Longevity. Ziyun Wu et al. Nov 2013.
293) Awad R, Levac D, Cybulska P, Merali Z, Trudeau VL, Arnason JT. Effects of traditionally used anxiolytic botanicals on enzymes of the gamma-aminobutyric acid (GABA) system. Can J Physiol Pharmacol. 2007;85(9):933-942.18066140/
294) Low-protein diets supplemented with glutamic acid or aspartic acid ameliorate intestinal damage in weaned piglets challenged with hydrogen peroxide. Shuai Chen, et al. Mar 2021.
295) Metabolic responses to prolonged fasting and subsequent refeeding in the pig. G J Klain et al. Mary 1977.
296) Combination of aspartic acid and glutamic acid inhibits tumor cell proliferation. Yoshie Yamaguchi. 2016.
297) Methuselah, a Bristlecone Pine is Thought to be the Oldest Living Organism on Earth. Robert Hudson Westover, Public Affairs Specialist, USDA Forest Service in Forestry. Apr 21, 2011.
298) Chronicling Climate Change. Julia Rothchild. Smithsonian Science Education Center. https://ssec.si.edu
299) Overexpression of SOX9 alleviates the progression of human

osteoarthritis in vitro and in vivo. Yuanming Ouyang et al. Aug 2019.
300) In Vitro Screening Studies on Eight Commercial Essential Oils-Derived Compounds to Identify Promising Natural Agents for the Prevention of Osteoporosis. Marta Trzaskowska et al. Apr 2023.
301) Fumigant and repellent activities of essential oil extracted from Artemisia dubia and its main compounds against two stored product pests. Jun-Yu Liang et al. May 2018.
302) Properties of Resveratrol: In Vitro and In Vivo Studies about Metabolism, Bioavailability, and Biological Effects in Animal Models and Humans. J. Gambini et al. June 2015.
303) Properties of Resveratrol: In Vitro and In Vivo Studies about Metabolism, Bioavailability, and Biological Effects in Animal Models and Humans. J. Gambini et al. June 2015.
304) Properties of Resveratrol: In Vitro and In Vivo Studies about Metabolism, Bioavailability, and Biological Effects in Animal Models and Humans. J. Gambini et al. June 2015.
305) Defense traits in the long-lived Great Basin bristlecone pine and resistance to the native herbivore mountain pine beetle. Barbara J. Bentz et al. Sept 2016.
306) Xylem Monoterpenes of Pines: Distribution, Variation, Genetics, Function. Richard H. Smith.
307) Antimicrobial Activity and Proposed Action Mechanism of 3-Carene against Brochothrix thermosphacta and Pseudomonas fluorescens. Huizhen Shu et al. Sept 2019.
308) Mountain Pine Beetles Use Volatile Cues to Locate Host Limber Pine and Avoid Non-Host Great Basin Bristlecone Pine. Curtis A Gray et al. Sept 2015.
309) Defense traits in the long-lived Great Basin bristlecone pine and resistance to the native herbivore mountain pine beetle. Barbara J. Bentz, et al. Jan 2017.
310) The Mode of Action of Cyclic Monoterpenes (−)-Limonene and (+)-α-Pinene on Bacterial Cells. Olga E. Melkina, et al. May 2021.
311) Therapeutic Potential of α- and β-Pinene: A Miracle Gift of Nature. Bahare Salehi, et al. Nov 2019.
312) A Review of the Potential Use of Pinene and Linalool as Terpene-Based Medicines for Brain Health: Discovering Novel Therapeutics in the Flavours and Fragrances of Cannabis. Katrina Weston-Green et al. Aug 2021.
313) A Review of the Potential Use of Pinene and Linalool as Terpene-Based Medicines for Brain Health: Discovering Novel Therapeutics in the Flavours and Fragrances of Cannabis. Katrina Weston-Green et al. Aug 2021.
314) A Review of the Potential Use of Pinene and Linalool as Terpene-Based Medicines for Brain Health: Discovering Novel Therapeutics in the Flavours and Fragrances of Cannabis. Katrina Weston-Green et al. Aug 2021.
315) A Review of the Potential Use of Pinene and Linalool as Terpene-Based Medicines for Brain Health: Discovering Novel Therapeutics in the Flavours and Fragrances of Cannabis. Katrina Weston-Green et al. Aug

2021.
316) A Review of the Potential Use of Pinene and Linalool as Terpene-Based Medicines for Brain Health: Discovering Novel Therapeutics in the Flavours and Fragrances of Cannabis. Katrina Weston-Green et al. Aug 2021.
317) Potential protective effects of alpha-pinene against cytotoxicity caused by aspirin in the IEC-6 cells. Hafsia Bouzenna et al. Sept 2017.
318) In vitro neuroprotective potential of the monoterpenes α-pinene and 1,8-cineole against H2O2-induced oxidative stress in PC12 cells. María Porres et al. 2016.
319) Analysis of the glutathione S-transferase (GST) gene family. Daniel W Nebert 1, Vasilis Vasiliou et al. 2004.
320) Attenuating effect of α-pinene on neurobehavioural deficit, oxidative damage and inflammatory response following focal ischaemic stroke in rat. Mahdieh Khoshnazar et al. Nov 2019.
321) In vitro assessment of cytogenetic and oxidative effects of α-pinene. Hasan Türkez 1, Elanur Aydın. Jan 2016.
322) A Review of the Potential Use of Pinene and Linalool as Terpene-Based Medicines for Brain Health: Discovering Novel Therapeutics in the Flavours and Fragrances of Cannabis. Katrina Weston-Green et al. Aug 2021.
323) A Review of the Potential Use of Pinene and Linalool as Terpene-Based Medicines for Brain Health: Discovering Novel Therapeutics in the Flavours and Fragrances of Cannabis. Katrina Weston-Green et al. Aug 2021.
324) Insights into flavor and key influencing factors of Maillard reaction products: A recent update. Shuyun Liu et al. Sept 2022.
325) Preliminary Study of Aromatic Components in Herbaceous Peonies of 'Yangfei Chuyu'and 'Dafugui. HUANG Xue et al. May 2010. ACTA HORTICULTURAE SINICA ›› 2010, Vol. 37 ›› Issue (5): 817-822.
326) Paeonia × suffruticosa (Moutan Peony)—A Review of the Chemical Composition, Traditional and Professional Use in Medicine. Halina Ekiert, et al. Dec 2022.
327) Antioxidant effect of peony seed oil on aging mice. Xiao-Miao Han et al. Dec 2017.
328) Eryngium foetidum L. (Apiaceae): A Literature Review of Traditional Uses, Chemical Composition, and Pharmacological Activities. Thiara L. M. Rodrigues et al. 2022.
329) Effects of Exogenous Ethylene and 1-Methylcyclopropene on Postharvest Physicochemical and Aroma Characteristics of 'Shengxin' Mango. LIANG Minhua et al. 2020.
330) Effect of Rootstock on Mango Fruit Susceptibility to Infestation byAnastrepha obliqua. A. VAZQUEZ-LUNA et al. 1991. Journal of Economic Entomology.
331) A Review of the Potential Use of Pinene and Linalool as Terpene-Based Medicines for Brain Health: Discovering Novel Therapeutics in the Flavours and Fragrances of Cannabis. Katrina Weston-Green et al. Aug 2021.
332) A Review of the Potential Use of Pinene and Linalool as Terpene-

Based Medicines for Brain Health: Discovering Novel Therapeutics in the Flavours and Fragrances of Cannabis. Katrina Weston-Green et al. Aug 2021.
333) A Review of the Potential Use of Pinene and Linalool as Terpene-Based Medicines for Brain Health: Discovering Novel Therapeutics in the Flavours and Fragrances of Cannabis. Katrina Weston-Green et al. Aug 2021.
334) A Review of the Potential Use of Pinene and Linalool as Terpene-Based Medicines for Brain Health: Discovering Novel Therapeutics in the Flavours and Fragrances of Cannabis. Katrina Weston-Green et al. Aug 2021.
335) Hypothalamic Menin regulates systemic aging and cognitive decline. Lige Leng et al. Mar 2023.
336) Roman chamomile inhalation combined with clomipramine treatment improves treatment-resistant depression-like behavior in mice. Narumi Hashikawa-Hobara et al. Oct 2019.
337) Cognitive Aging and the Hippocampus in Older Adults. Andrew O'Shea1, et al. Dec 2016.
338) A Review of the Potential Use of Pinene and Linalool as Terpene-Based Medicines for Brain Health: Discovering Novel Therapeutics in the Flavours and Fragrances of Cannabis. Katrina Weston-Green et al. Aug 2021.
339) A Review of the Potential Use of Pinene and Linalool as Terpene-Based Medicines for Brain Health: Discovering Novel Therapeutics in the Flavours and Fragrances of Cannabis. Katrina Weston-Green et al. Aug 2021.
340) A Review of the Potential Use of Pinene and Linalool as Terpene-Based Medicines for Brain Health: Discovering Novel Therapeutics in the Flavours and Fragrances of Cannabis. Katrina Weston-Green et al. Aug 2021.
341) Effects of Controlled Atmosphere on the Storage Quality and Aroma Compounds of Lemon Fruits Using the Designed Automatic Control Apparatus. Yuan Ma, et al. Jun 2019.
342) A Review of the Potential Use of Pinene and Linalool as Terpene-Based Medicines for Brain Health: Discovering Novel Therapeutics in the Flavours and Fragrances of Cannabis. Katrina Weston-Green et al. Aug 2021.
343) Citrus Waste as Source of Bioactive Compounds: Extraction and Utilization in Health and Food Industry. Zahra Maqbool et al. Feb 2023.
344) A Review of the Potential Use of Pinene and Linalool as Terpene-Based Medicines for Brain Health: Discovering Novel Therapeutics in the Flavours and Fragrances of Cannabis. Katrina Weston-Green et al. Aug 2021.
345) A Review of the Potential Use of Pinene and Linalool as Terpene-Based Medicines for Brain Health: Discovering Novel Therapeutics in the Flavours and Fragrances of Cannabis. Katrina Weston-Green et al. Aug 2021.
346) A Review of the Potential Use of Pinene and Linalool as Terpene-Based Medicines for Brain Health: Discovering Novel Therapeutics in the

Flavours and Fragrances of Cannabis. Katrina Weston-Green et al. Aug 2021.

347) Effect of 4 weeks of frankincense consumption on explicit motor memory and serum BDNF in elderly men. Elham ASADI et al. Aug 2019.

348) Effects of Frankincense Compounds on Infection, Inflammation, and Oral Health. Cássio Luiz Coutinho Almeida-da-Silva et al. June 2022.

349) Effects of essential oils on central nervous system: Focus on mental health. Lorena R. and Lizarraga-Valderrama. Aug 2020.

350) Great Basin Bristlecone Pine Volatiles as a Climate Change Signal Across Environmental Gradients. Curtis A. Gray et al. Apr 2019.

351) A Review of the Potential Use of Pinene and Linalool as Terpene-Based Medicines for Brain Health: Discovering Novel Therapeutics in the Flavours and Fragrances of Cannabis. Katrina Weston-Green et al. Aug 2021.

352) Combined Analysis of Volatile Terpenoid Metabolism and Transcriptome Reveals Transcription Factors Related to Terpene Synthase in Two Cultivars of Dendrobium officinale Flowers. Ninghong Li et al. Apr 2021.

353) Preparation of Dendrobium officinale Flower Anthocyanin and Extended Lifespan in Caenorhabditis elegans. Shuangxi Li et al. Dec 2022.

354) Anthocyanins: Metabolic Digestion, Bioavailability, Therapeutic Effects, Current Pharmaceutical/Industrial Use, and Innovation Potential. Huseyin Ayvaz, et al. Jan 2023.

355) A Review of Factors Affecting Anthocyanin Bioavailability: Possible Implications for the Inter-Individual Variability. Merve Eda Eker et al. Dec 2019.

356) Effects of In Vitro Gastrointestinal Digestion on the Antioxidant Capacity and Anthocyanin Content of Cornelian Cherry Fruit Extract. Luminita David, et al. May 2019.

357) Roles and Mechanisms of Hawthorn and Its Extracts on Atherosclerosis: A Review. Min Wu, et al. Feb 2020.

358) The Therapeutic Potential of Rosemary (Rosmarinus officinalis) Diterpenes for Alzheimer's Disease. Solomon Habtemariam. Jan 2016.

359) The investigation of some bioactive compounds and antioxidant properties of hawthorn (Crataegus monogyna subsp. monogyna Jacq). Serhat Keser et al. Apr 2014.

360) Chemical Properties Of Hawthorn (Crataegus L. Spp.) Taxa Naturally Distributed In Western Anatolia Partof Turkey. Sevgin Özdern et al. 2016.

361) Effects of food formulation and thermal processing on flavones in celery and chamomile. Gregory L. Hostetler, et al. Apr 2013.

362) Effects of food formulation and thermal processing on flavones in celery and chamomile. Gregory L. Hostetler, et al. Apr 2013.

363) Effects of food formulation and thermal processing on flavones in celery and chamomile. Gregory L. Hostetler, et al. Apr 2013.

364) Effects of food formulation and thermal processing on flavones in celery and chamomile. Gregory L. Hostetler, et al. Apr 2013.

365) Anti-Inflammatory Effect of an Apigenin-Maillard reaction Product in Macrophages and Macrophage-Endothelial Cocultures. Qian Zhou et al.

May 2019.
366) Apigenin glycosides from green pepper enhance longevity and stress resistance in Caenorhabditis elegans. Abdeen E Elkhedir et al. June 2022.
367) The Influence Of Polyphenols On The Formation Of Free Radicals Detected In Maillard reaction Model Systems. Marijana B. Sakač et al. University of Novi Sad, Institute of Food Technology, 21000 Novi Sad, Bulevar cara Lazara 1, Serbi. DOI: 10.5937/FFR1802187S. UDK 637.54'65:664.8.037.5:536.621/.626.
368) Polyphenols and Maillard reaction Products in Dried Prunus spinosa Fruits: Quality Aspects and Contribution to Anti-Inflammatory and Antioxidant Activity in Human Immune Cells Ex. Anna Magiera et al. May 2022.
369) Wang N., Jiang S., Zhang Z., Fang H., Xu H., Wang Y., Chen X. Malus sieversii: The origin, flavonoid synthesis mechanism, and breeding of red-skinned and red-fleshed apples. Hortic. Res. 2018;5:1–12. doi: 10.1038/s41438-018-0084-4.
370) Effects of Different Drying Methods on the Retention of Bioactive Compounds, On-Line Antioxidant Capacity and Color of the Novel Snack from Red-Fleshed Apples. Aneta Wojdyło et al. Nov 2020.
371) Headspace and essential oil analysis of apple flowers. Gerhard. Buchbauer et al. Jan 1993.
372) Biochemistry of Apple Aroma: A Review. Miguel Espino-Díaz et al. Dec 2016.
373) Food polyphenols and Maillard reaction: regulation effect and chemical mechanism. Zisheng Han et al. Nov 2022.
374) Can Dietary Polyphenols Prevent the Formation of Toxic Compounds from Maillard reaction?. Serena Del Turco, Giuseppina Basta. 2016.
375) Comprehensive characterization of lotus root polysaccharide-phenol complexes. Yang Yi et al. Jan 2022.
376) Antioxidant and hepatic protective effects of lotus root hot water extract with taurine supplementation in rats fed a high fat diet. Huan Du, et al. Aug 2010.
377) Anti-Inflammatory Effects of Fermented Lotus Root and Linoleic Acid in Lipopolysaccharide-Induced RAW 264.7 Cells. Sung Min Kim et al. Nov 2020
378) Lotus root extract inhibits skin damage through suppression of collagenase production in vitro. Akira Iwamoto et al. Apr 2022.
379) Polyphenols extract from lotus seedpod (Nelumbo nucifera Gaertn.): Phenolic compositions, antioxidant, and antiproliferative activities. Yingbin Shen et al. Apr 2019.
380) Shared by Nature: A thousand-year-old lotus burst into bloom. CGTN. www.news.cgtn.com
381) Long-living lotus: germination and soil {gamma}-irradiation of centuries-old fruits, and cultivation, growth, and phenotypic abnormalities of offspring. J Shen-Miller et al. Feb 2002.
382) Editorial: Chemical and biological changes of polyphenols caused by food thermal processing. Ying Zhang and Huilin Liu. July 2022.

383) Hydrolyzable tannins from the cores of Cornus officinalis with inhibitory activity in vitro on the formation of advanced glycation end products (AGEs). GY Lee et al 2008.
384) Prevention of Enzymatic Browning by Natural Extracts and Genome-Editing: A Review on Recent Progress. Norfadilah Hamdan et al. Feb 2022.
385) Red-skinned onion phenolic compounds stability and bioaccessibility: A comparative study between deep-frying and air-frying. Alice Cattivelli et al. Jan 2023.
386) Anti-Inflammatory Effect of an Apigenin-Maillard reaction Product in Macrophages and Macrophage-Endothelial Cocultures. Qian Zhou et al. May 2019.
387) Potential Synergistic Supplementation of NAD+ Promoting Compounds as a Strategy for Increasing Healthspan. Arastu Sharma et al. Jan 2023.
388) Apigenin and its methylglyoxal-adduct inhibit advanced glycation end products-induced oxidative stress and inflammation in endothelial cells. Qian Zhou et al. Aug 2019.
389) Potential Synergistic Supplementation of NAD+ Promoting Compounds as a Strategy for Increasing Healthspan. Arastu Sharma et al. Jan 2023.
390) Healthy Lifestyle Recommendations: Do the Beneficial Effects Originate from NAD+ Amount at the Cellular Level?. Borut Poljsak et al. Dec 2020.
391) Potential Synergistic Supplementation of NAD+ Promoting Compounds as a Strategy for Increasing Healthspan. Arastu Sharma et al. Jan 2023.
392) Anti-Inflammatory Effect of an Apigenin-Maillard reaction Product in Macrophages and Macrophage-Endothelial Cocultures. Qian Zhou et al. May 2019.
393) Effects of food formulation and thermal processing on flavones in celery and chamomile. Gregory L. Hostetler, et al. Apr 2013.
394) Effects of food formulation and thermal processing on flavones in celery and chamomile. Gregory L. Hostetler, et al. Apr 2013.
395) USDA Database for the Flavonoid Content of Selected Foods. Release 3.1 Prepared by Seema Bhagwat, David B. Haytowitz, and Joanne M. Holden (ret.) Nutrient Data Laboratory. Beltsville Human Nutrition Research Center. Agricultural Research Service. U.S. Department of Agriculture. December 2013. Slightly revised, May 2014.
396) Anthocyanins: Metabolic Digestion, Bioavailability, Therapeutic Effects, Current Pharmaceutical/Industrial Use, and Innovation Potential. Huseyin Ayvaz, et al. Jan 2023.
397) Research Progress on Absorption, Metabolism, and Biological Activities of Anthocyanins in Berries: A Review. Hongkun Xue et al. Dec 2022.
398) Anthocyanins and Vascular Health: A Matter of Metabolites. Joseph Festa et al. Apr 2023.
399) Bioavailability of anthocyanins. Jim Fang et al. 2014.
400) Mulberry anthocyanins, cyanidin 3-glucoside and cyanidin 3-

rutinoside, increase the quantity of mitochondria during brown adipogenesis. Yilin You et al. Sept 2017.

401) Mulberry anthocyanins, cyanidin 3-glucoside and cyanidin 3-rutinoside, increase the quantity of mitochondria during brown adipogenesis. Yilin You et al. Sept 2017.

402) Anthocyanins, Microbiome and Health Benefits in Aging. Rachel Hair et al. Feb 2021.

403) Anthocyanins, Microbiome and Health Benefits in Aging. Rachel Hair et al. Feb 2021.

404) Antioxidant and alpha-glucosidase inhibitory activity of colored grains in China. Yang Yao et al. Jan 2010.

405) Impact of Polyphenolic-Food on Longevity: An Elixir of Life. An Overview. Rosaria Meccariello and Stefania D'Angelo et al. Apr 2021.

406) Sweeteners'Influence on In Vitro α-Glucosidase Inhibitory Activity, Cytotoxicity, Stability and In Vivo Bioavailability of the Anthocyanins from Lingonberry Jams. Teodora Scrob et al. June 2023.

407) A Review of Factors Affecting Anthocyanin Bioavailability: Possible Implications for the Inter-Individual Variability. Merve Eda Eker et al. Dec 2019.

408) Effects of In Vitro Gastrointestinal Digestion on the Antioxidant Capacity and Anthocyanin Content of Cornelian Cherry Fruit Extract. Luminita David, et al. May 2019.

409) Effects of In Vitro Gastrointestinal Digestion on the Antioxidant Capacity and Anthocyanin Content of Cornelian Cherry Fruit Extract. Luminita David, et al. May 2019.

410) Effects of In Vitro Gastrointestinal Digestion on the Antioxidant Capacity and Anthocyanin Content of Cornelian Cherry Fruit Extract. Luminita David, et al. May 2019.

411) Anthocyanins: Factors Affecting Their Stability and Degradation. Bianca Enaru et al. Dec 2021.

412) Pelargonidin exhibits restoring effects against amyloid β-induced deficits in the hippocampus of male rats. Sara Soleimani Asl et al. Dec 2019.

413) Pharmacological Activities and Chemical Stability of Natural and Enzymatically Acylated Anthocyanins: A Comparative Review. Jimena Yañez-Apam et al. Apr 2023.

414) Increase in thermal stability of strawberry anthocyanins with amino acid copigmentation. Alev Bingöl et al. Aug 2022.

415) Goji Berries: Valuable Sources of Nutrients and Bioactive Compounds. Bojana B. Vidović et al. Food Science, 2023.

416) Bioactivities and Microbial Quality of Lycium Fruits (Goji) Extracts Derived by Various Solvents and Green Extraction Methods. Katarzyna Rajkowska, et al. Nov 2022.

417) Stability and bioaccessibility of anthocyanins in bakery products enriched with anthocyanins. Sibel Karakaya. et al. Aug 2016.

418) Anthocyanins and Human Health: An In Vitro Investigative Approach. Mary Ann Lila. Dec 2004.

419) Stimulatory effect of cyanidin 3-glycosides on the regeneration of rhodopsin. Hitoshi Matsumoto et al. Jun 2003.

420) pH-dependent interaction of rhodopsin with cyanidin-3-glucoside.

2. Functional aspects. Kalyan C Tirupula et al. Mar 2009.
421) Gut and microbial resveratrol metabolite profiling after moderate long-term consumption of red wine versus dealcoholized red wine in humans by an optimized ultra-high-pressure liquid chromatography tandem mass spectrometry method. Maria Rotches-Ribalta et al. Nov 2012.
422) Effect of chlorogenic acid on hydroxymethylfurfural in different Maillard reaction systems. S.S. Jiang et al. Jan 2013.
423) A Review of Factors Affecting Anthocyanin Bioavailability: Possible Implications for the Inter-Individual Variability. Merve Eda Eker et al. Dec 2019.
424) Research Progress on Absorption, Metabolism, and Biological Activities of Anthocyanins in Berries: A Review. Hongkun Xue et al. Dec 2022.
425) Malvidin . www.phenol-explorer.eu
426) Food Anthocyanins: Malvidin and Its Glycosides as Promising Antioxidant and Anti-Inflammatory Agents with Potential Health Benefits. Anna Merecz-Sadowska Jul 2023.
427) Malvidin-3-glucoside bioavailability in humans after ingestion of red wine, dealcoholized red wine and red grape juice. A Bub et al. Jun 2001.
428) Anthocyanins Are Efficiently Absorbed from the Stomach in Anesthetized Rats. Séverine Talavéra et al. Dec 2003.
429) Saskatoon and Wild Blueberries Have Higher Anthocyanin Contents than Other Manitoba Berries. Farah S. Hosseinian and Trust Beta. Dec 2007.
430) Malvidin-3-glucoside bioavailability in humans after ingestion of red wine, dealcoholized red wine and red grape juice. A Bub et al. Jun 2001.
431) Stability of Anthocyanins and Their Degradation Products from Cabernet Sauvignon Red Wine under Gastrointestinal pH and Temperature Conditions. by Ping Yang et al. Feb 2018.
432) Ingestion of red wine significantly increases plasma phenolic acid concentrations but does not acutely affect ex vivo lipoprotein oxidizability. Rima Abu-Amsha Caccetta et al. Jan 2000.
433) Anti-glycative and anti-inflammatory effects of protocatechuic acid in brain of mice treated by d-galactose. Shih-jei Tsai a b, Mei-chin Yin et al. Sept 2012.
434) State of the Art of Anthocyanins: Antioxidant Activity, Sources, Bioavailability, and Therapeutic Effect in Human Health. Noelia Tena et al. May 2020.
435) Aglycones and Sugar Moieties Alter Anthocyanin Absorption and Metabolism after Berry Consumption in Weanling Pigs. Xianli Wu et al. Oct 2005.
436) Anthocyanins and Vascular Health: A Matter of Metabolites. Joseph Festa et al. Apr 2023.
437) Anthocyanins and Vascular Health: A Matter of Metabolites. Joseph Festa et al. Apr 2023.
438) Harrison DE, Strong R, Alavez S, et al. Acarbose improves health and lifespan in aging HET3 mice. Aging Cell. 2019;18:e12898. doi: 10.1111/acel.12898.
439) Antiaging agents: safe interventions to slow aging and healthy life

span extension. Ji-Kai Liu. May 2022.
440) Anthocyanins and Vascular Health: A Matter of Metabolites. Joseph Festa et al. Apr 2023.
441) Anthocyanins and Vascular Health: A Matter of Metabolites. Joseph Festa et al. Apr 2023.
442) Aglycones and Sugar Moieties Alter Anthocyanin Absorption and Metabolism after Berry Consumption in Weanling Pigs. Xianli Wu et al. Oct 2005.
443) Bioavailability and molecular activities of anthocyanins as modulators of endothelial function. Antonio Speciale et al. May 2014.
444) Anthocyanins and Vascular Health: A Matter of Metabolites. Joseph Festa et al. Apr 2023.
445) Impact of Cyanidin-3-Glucoside on Gut Microbiota and Relationship with Metabolism and Inflammation in High Fat-High Sucrose Diet-Induced Insulin Resistant Mice. Fei Huang. July 2020.
446) The safety and pharmacokinetics of cyanidin-3-glucoside after 2-week administration of black bean seed coat extract in healthy subjects. Sangil Jeon et al. Aug 2012.
447) The Safety and Pharmacokinetics of Cyanidin-3-Glucoside after 2-Week Administration of Black Bean Seed Coat Extract in Healthy Subjects. Sangil Jeon, et al. Aug 2012.
448) A Systematic Review of Antiaging Effects of 23 Traditional Chinese Medicines. Lixin Wang et al. May 2021.
449) Antioxidant activity and absorption of cyanidin-3-O-glucoside liposomes in GES-1 cells in vitro. Zhao Quan et al. Mar 2020.
450) Elderberry (Sambucus nigra L.) Wine: A Product Rich in Health Promoting Compounds. Valentina Schmitzer et al. Sept 2010.
451) Advancement of Protein and Polysaccharide-Based Biopolymers for Anthocyanin Encapsulation. Jiahui Song et al. June 2022.
452) Combined anthocyanins and bromelain supplement improves endothelial function and skeletal muscle oxygenation status in adults: a double-blind placebo-controlled randomised crossover clinical tria. Levers, Dalton and Galvan. Jul 2020.
453) Wikipeida. Polyphenol.
454) Impact of Polyphenolic-Food on Longevity: An Elixir of Life. An Overview. Rosaria Meccariello and Stefania D'Angelo et al. Apr 2021.
455) The Reciprocal Interactions between Polyphenols and Gut Microbiota and Effects on Bioaccessibility. Tugba Ozdal et al. Feb 2016.
456) Beneficial Effects of Dietary Polyphenols on Gut Microbiota and Strategies to Improve Delivery Efficiency. Amit Kumar Singh et al. Sept 2019.
457) Beneficial Effects of Dietary Polyphenols on Gut Microbiota and Strategies to Improve Delivery Efficiency. Amit Kumar Singh et al. Sept 2019.
458) Impact of Polyphenolic-Food on Longevity: An Elixir of Life. An Overview. Rosaria Meccariello and Stefania D'Angelo et al. Apr 2021.
459) Colonic availability of polyphenols and D-(-)-quinic acid after apple smoothie consumption. Stephanie Hagl et al. Mar 2011.
460) New Smoothie Products Based on Pumpkin, Banana, and Purple

Carrot as a Source of Bioactive Compounds. Marcin Kidoń and Pascaline Aimee Uwineza et al. May 2022.
461) In vitro gastrointestinal evaluation of a juçara-based smoothie: effect of processing on phenolic compounds bioaccessibility. Leilson de Oliveira Ribeiro et al. Aug 2019.
462) New Insight on Phenolic Composition and Evaluation of the Vitamin C and Nutritional Value of Smoothies Sold on the Spanish Market. María del Carmen Razola-Díaz et al. Nov 2022.
463) Uptake and bioavailability of anthocyanins and phenolic acids from grape/blueberry juice and smoothie in vitro and in vivo. Sabine Kuntz et al. Apr 2015.
464) The smell of longevity: a combination of Volatile Organic Compounds (VOCs) can discriminate centenarians and their offspring from age-matched subjects and young controls. Maria Conte et al. Dec 2019.
465) Impact of Polyphenolic-Food on Longevity: An Elixir of Life. An Overview. Rosaria Meccariello and Stefania D'Angelo et al. Apr 2021.
466) Influence of red wine polyphenols and ethanol on the gut microbiota ecology and biochemical biomarkers. María Isabel Queipo-Ortuño et al. May 2012.
467) If you want to boost immunity, look to the gut. Sandy Cohen. March 19, 2021. www.uclahealth.org
468) Anthocyanins, Microbiome and Health Benefits in Aging. Rachel Hair et al. Feb 2021.
469) Anthocyanins, Microbiome and Health Benefits in Aging. Rachel Hair et al. Feb 2021.
470) Formation of short-chain Fatty acids, excretion of anthocyanins, and microbial diversity in rats fed blackcurrants, blackberries, and raspberries. Greta Jakobsdottir et al. June 2013.
471) Undigested glycated lentil proteins modulate the gut microbiota profile but not the metabolites in vitro. Ruth T. Boachie et al. Aug 2023.
472) Impact of Edible Cricket Consumption on Gut Microbiota in Healthy Adults, a Double-blind, Randomized Crossover Trial. Valerie J Stull et al. Jul 2018.
473) Effect of dark sweet cherry powder consumption on the gut microbiota, short-chain fatty acids, and biomarkers of gut health in obese db/db mice. Jose F. Garcia-Mazcorro et al. Jan 2018.
474) Daily ingestion of Akkermansia mucciniphila for one month promotes healthy aging and increases lifespan in old female mice. Estefanía Díaz-Del Cerro et al. Feb 2022.
475) Metabolome and gut microbiota variation with long-term intake of Panax ginseng extracts on rats. Yifan Sun. 2018.
476) Evaluation of the Quality of Yogurt Using Ginseng Extract Powder and Probiotic Lactobacillus plantarum NK181. Hye Ji Jang et al. Dec 2018.
477) Co-Fermentation by Lactobacillus brevis B7 Improves the Antioxidant and Immunomodulatory Activities of Hydroponic Ginseng-Fortified Yogurt. Myung-Wook Song et al. Sept 2021.
478) Physicochemical Characteristics and Antioxidant Capacity in Yogurt Fortified with Red Ginseng Extract. Jieun Jung, et al. June 2016.
479) Ginseng volatile oil prolongs the lifespan and healthspan of

Caenorhabditis elegans. Lixin Wang et al. Aug 2022.
480) Ginseng volatile oil prolongs the lifespan and healthspan of Caenorhabditis elegans. Lixin Wang et al. Aug 2022.
481) Screening of promising chemotherapeutic candidates from plants extracts. Junei Kinjo et al. Apr 2016.
482) Volatile Compositions of Panax ginseng and Panax quinquifolium Grown for Different Cultivation Years. Yejin Kim et al. Dec 2022.
483) Anthocyanins, Microbiome and Health Benefits in Aging. Rachel Hair et al. Feb 2021.
484) Anthocyanins, Microbiome and Health Benefits in Aging. Rachel Hair et al. Feb 2021.
485) Biological Activities of Grape Seed By-Products and Their Potential Use as Natural Sources of Food Additives in the Production of Balady Bread. Haiam O. Elkatry, et al. June 2022.
486) Kinetics of thermal modifications in a grape seed extract. Gabriel Davidov-Pardo et al. Jul 2011.
487) In Vitro Remineralization Effects of Grape Seed Extract on Artificial Root Caries. Qian Xie et al. Sept 2008.
488) Use of Grape Pomace Phenolics to Counteract Endogenous and Exogenous Formation of Advanced Glycation End-Products. Pedapati S. C. et al. Aug 2019.
489) Use of Grape Pomace Phenolics to Counteract Endogenous and Exogenous Formation of Advanced Glycation End-Products. Pedapati S. C. et al. Aug 2019.
490) Antioxidant and antiglycation activity of selected dietary polyphenols in a cookie model. Xinchen Zhang et al. Feb 2014.
491) Use of Grape Pomace Phenolics to Counteract Endogenous and Exogenous Formation of Advanced Glycation End-Products. Pedapati S. C. et al. Aug 2019.
492) Use of Grape Pomace Phenolics to Counteract Endogenous and Exogenous Formation of Advanced Glycation End-Products. Pedapati S. C. et al. Aug 2019.
493) Biological Activities of Grape Seed By-Products and Their Potential Use as Natural Sources of Food Additives in the Production of Balady Bread. Haiam O. Elkatry, et al. June 2022.
494) Use of Grape Pomace Phenolics to Counteract Endogenous and Exogenous Formation of Advanced Glycation End-Products. Pedapati S. C. et al. Aug 2019.
495) Physical, microbiological and rheological properties of probiotic yogurt supplemented with grape extract. Denise Felix da Silva et al. Mar 2017.
496) Physical, microbiological and rheological properties of probiotic yogurt supplemented with grape extract. Denise Felix da Silva et al. Mar 2017.
497) Preparation and evaluation of yogurt fortified with probiotics jelly candy enriched with grape seeds extract nanoemulsion. Samah M. El-Sayed et al. Apr 2022.
498) Natural nutraceuticals for enhancing yogurt properties: a review. Ahmed K. Rashwan et al. March 2023.

499) Natural nutraceuticals for enhancing yogurt properties: a review. Ahmed K. Rashwan et al. March 2023.
500) Natural nutraceuticals for enhancing yogurt properties: a review. Ahmed K. Rashwan et al. March 2023.
501) Natural nutraceuticals for enhancing yogurt properties: a review. Ahmed K. Rashwan et al. March 2023.
502) Chamomile (Matricaria chamomilla. A Review of Ethnomedicinal Use, Phytochemistry and Pharmacological Uses. Amina El Mihyaoui et al. Mar 2022.
503) Comparative Analysis of Phenolic Composition of Six Commercially Available Chamomile (Matricaria chamomilla L.) Extracts: Potential Biological Implications. Maria Valeria Catani et al. Sept 2021.
504) Effect of drying methods on yield, physicochemical properties, and total polyphenol content of chamomile extract powder. www.frontiersin.org. Sin Yee Lee et al. Nov 2020.
505) Antioxidant and anticancer activities of chamomile (Matricaria recutita L.). Bayan Al-Dabbagh et al. Jan 2019.
506) Antioxidant and anticancer activities of chamomile (Matricaria recutita L.). Bayan Al-Dabbagh et al. Jan 2019.
507) Natural nutraceuticals for enhancing yogurt properties: a review. Ahmed K. Rashwan et al. March 2023.
508) Natural nutraceuticals for enhancing yogurt properties: a review. Ahmed K. Rashwan et al. March 2023.
509) Natural nutraceuticals for enhancing yogurt properties: a review. Ahmed K. Rashwan et al. March 2023.
510) Mechanisms of protection against damage mediated by the Maillard reaction in aging. V M Monnier et al. 1991.
511) Lysine-glucose Maillard reaction products promote longevity and stress tolerance in Caenorhabditis elegans via the insulin/IGF-1 signaling pathway, Issei YokoyamaIssei et al. Dec 2021.
512) Coffee Extends Yeast Chronological Lifespan through Antioxidant. Jadwiga Czachor et al. Dec 2020.
513) Assessing polyphenols content and antioxidant activity in coffee beans according to origin and the degree of roasting. Ewa Dybkowska et al. 2017.
514) The Argus (Melbourne, Vic. : 1848 - 1957) Sat 30 Jun 1934. Page 21 ZARO AGHA. Dead at 160 years. World's Oldest Man.
515) Zaro Agha Dies At Mooted Age Of Ighears Scientists To Test Longevity Claim Of Turk By Autopsy. San Pedro News Pilot, Volume 7, Number 100, 29 June 1934.
516) Red-skinned onion phenolic compounds stability and bioaccessibility: A comparative study between deep-frying and air-frying. Alice Cattivelli, et al. Jan 2023.
517) Air frying combined with grape seed extract inhibits Nε-carboxymethyllysine and Nε-carboxyethyllysine by controlling oxidation and glycosylation. Zongshuai Zhu et al. Dec 2020.
518) Effects of amino acids on the physiochemical properties of potato starch. Min Cui et al. May 2014.
519) Novel 1,2,4-Triazolo[1,5-a]pyridines and Their Fused Ring Systems

Attenuate Oxidative Stress and Prolong Lifespan of Caenorhabiditis elegans. Ramadan Ahmed Mekheimer et al. Apr 2012.

520) The volatile constituents of roasted coffee. E. B. Hughes and R. F. Smith. Nov 1949.

521) Novel 1,2,4-Triazolo[1,5-a]pyridines and Their Fused Ring Systems Attenuate Oxidative Stress and Prolong Lifespan of Caenorhabiditis elegans. Ramadan Ahmed Mekheimer et al. Apr 2012.

522) Control of Maillard reactions in Foods: Strategies and Chemical Mechanisms. Marianne N. Lund*†‡Orcid and Colin A. Ray et al. May 2017.

523) An Investigation of the Complexity of Maillard reaction Product Profiles from the Thermal Reaction of Amino Acids with Sucrose Using High Resolution Mass Spectrometry. Agnieszka Golon et al. Sept 2014.

524) Sucrose degradation upon heating and the influence of amino acids on pyrolysis. I A Egorov et al Sept 1974.

525) Role of Chlorogenic Acids in Controlling Oxidative and Inflammatory Stress Conditions. Ningjian Liang and David D. Kitts et al. Dec 2015.

526) Synergistic Effects of Polyphenols and Methylxanthines with Leucine on AMPK/Sirtuin-Mediated Metabolism in Muscle Cells and Adipocytes. Antje Bruckbauer and Michael B. Zemel. Feb 2014.

527) Role of Chlorogenic Acids in Controlling Oxidative and Inflammatory Stress Conditions. Ningjian Liang and David D. Kitts et al. Dec 2015.

528) Quantification of Caffeoylquinic Acids in Coffee Brews by HPLC-DAD. Marzieh Moeenfard et al. Dec 2014.

529) Role of Chlorogenic Acids in Controlling Oxidative and Inflammatory Stress Conditions. Ningjian Liang and David D. Kitts et al. Dec 2015.

530) Chlorogenic Acid Activates Nrf2/SKN-1 and Prolongs the Lifespan of Caenorhabditis elegans via the Akt-FOXO3/DAF16a-DDB1 Pathway and Activation of DAF16f. Ferbian Milas Siswanto et al. Aug 2022.

531) Applying chlorogenic acid in an alginate scaffold of chondrocytes can improve the repair of damaged articular cartilage.. Xin Cheng et al. Apr 2018.

532) In vitro transformation of chlorogenic acid by human gut microbiota. Francisco Tomas-Barberan et al. May 2014.

533) Chlorogenic Acid Bioavailability Largely Depends on Its Metabolism by the Gut Microflora in Rats. Marie-Paule Gonthier et al. June 2003.

534) TAC-101, a benzoic acid derivative, inhibits liver metastasis of human gastrointestinal cancer and prolongs the life-span. Koji Murakami et al. May 1998.

535) 4-[3,5-Bis(trimethylsilyl)benzamido] benzoic acid (TAC-101) inhibits the intrahepatic spread of hepatocellular carcinoma and prolongs the life-span of tumor-bearing animals. K Murakam et al. Oct 1998.

536) Therapeutic potential of pyrrole and pyrrolidine analogs: an update. N. Jeelan Basha et al. Jan 2022.

537) Therapeutic potential of pyrrole and pyrrolidine analogs: an update. N. Jeelan Basha et al. Jan 2022.

538) Effects and Mechanisms of Resveratrol on Aging and Age-Related Diseases. Dan-Dan Zhou, et al. Jul 2021.

539) Resveratrol supplementation, where are we now and where should we go?. Marta G. Novelle, et al. Jan 2015.

540) Effect of food processing on antioxidants, their bioavailability and potential relevance to human health. Gamze Toydemir et al. June 2022.

541) Effect of food processing on antioxidants, their bioavailability and potential relevance to human health. Gamze Toydemir et al. June 2022.

542) The fever paradox. Linda Geddes Apr 2020.

543) Li H, Roxo M, Cheng X, Zhang S, Cheng H, Wink M. Pro-oxidant and lifespan extension effects of caffeine and related methylxanthines in Caenorhabditis elegans. Food Chem X. 2019;1:100005. doi: 10.1016/j.fochx.2019.100005.

544) Lublin A, Isoda F, Patel H, Yen K, Nguyen L, Hajje D, Schwartz M, Mobbs C. FDA-approved drugs that protect mammalian neurons from glucose toxicity slow aging dependent on cbp and protect against proteotoxicity. PLoS ONE. 2011;6(11):e27762.

545) Bridi JC, Barros AG, Sampaio LR, Ferreira JC, Antunes Soares FA, Romano-Silva MA. Lifespan extension induced by caffeine in Caenorhabditis elegans is partially dependent on adenosine signaling. Front Aging Neurosci. 2015;7:220. doi: 10.3389/fnagi.2015.00220.

546) Cyclodipeptides: An Overview of Their Biosynthesis and Biological Activity. Awdhesh Kumar Mishra et al. Oct 2017.

547) Impact of roasting on the phenolic and volatile compounds in coffee beans. Hanjing Wu et al. Jul 2022.

548) Salicylic Acid. www.pubchem.ncbi.nlm.nih.gov

549) Lagoecia cuminoides L., its antioxidant activity and polyphenolic constituents from Iran. Ateteh Bahmanzadegan et al. Aug 2019.

550) Salicylic Acid Content of Spices and Its Implications. Loraine Welch. December 2016.

551) Quantification of Chlorogenic Acid and Vanillin from Coffee Peel Extract and its Effect on α-Amylase Activity, Immunoregulation, Mitochondrial Oxidative Stress, and Tumor Suppressor Gene Expression Levels in H2O2-Induced Human Mesenchymal Stem Cells. Heba Khalil Alyahya et al. Nov 2021.

552) A salicylic acid derivative extends the lifespan of Caenorhabditis elegans by activating autophagy and the mitochondrial unfolded protein response. Mehrnaz Shamalnasab et al. Sept 2018.

553) Effect of dietary salicylic acid supplementation on performance and blood metabolites of sows and their litters. Serge Muhizi et al. Jul 2022.

554) Increased salicylate concentrations in urine of human volunteers after consumption of cranberry juice. Garry G Duthie et al. Apr 2005.

555) The Antioxidant Content of Coffee and Its In Vitro Activity as an Effect of Its Production Method and Roasting and Brewing Time. Maciej Górecki and Ewelina Hallmann. Apr 2020.

556) The Antioxidant Content of Coffee and Its In Vitro Activity as an Effect of Its Production Method and Roasting and Brewing Time. Maciej Górecki and Ewelina Hallmann. Apr 2020.

557) Deglycosylation is a key step in biotransformation and lifespan

effects of quercetin-3-O-glucoside in Caenorhabditis elegans. Montserrat Dueñas et al. Oct 2013.
558) Effects of thermal treatments on 10 major phenolics and their antioxidant contributions in Acer truncatum leaves and flowers. Lingguang Yang et al. June 2018.
559) Quercetin 3-O-glucoside. www.http://phenol-explorer.eu.
560) Effects of Cooking Method on the Antioxidant Activity and Inhibition of Lipid Peroxidation of the Javanese Salad "Pecel" Vegetables and Its Peanut Sauce Dressing. Gregorius Tsiompah et al. Feb 2021.
561) Effects of Cooking Method on the Antioxidant Activity and Inhibition of Lipid Peroxidation of the Javanese Salad "Pecel" Vegetables and Its Peanut Sauce Dressing. Gregorius Tsiompah et al. Feb 2021.
562) Nutritional and metabolic profiling of the globe artichoke (Cynara scolymus L. 'Capuanella' heads) in province of Caserta, Italy. Roberta Dosi et al. 2013.
563) Azizah A.H., Wee K.C., Azizah O., Azizah M. Effect of boiling and stir frying on total phenolics, carotenoids and radical scavenging activity of pumpkin (Cucurbita moschato) Int. Food Res. J. 2009;16:45–51.
564) Effect of Different Cooking Methods on Polyphenols, Carotenoids and Antioxidant Activities of Selected Edible Leaves. K. D. Prasanna P. Gunathilake et al. Sept 2018.
565) Effects of various heating methods on glucosinolate, carotenoid and tocopherol concentrations in broccoli. Eun-Sun Hwang and Gun-Hee Kim. Feb 2013.
566) Effect of different cooking methods on the content of vitamins and true retention in selected vegetables. Seongeung Lee et al. Apr 2018.
567) Effect of food processing on antioxidants, their bioavailability and potential relevance to human health. Gamze Toydemir et al. June 2022.
568) Extraction, Purification, and Hydrolysis Behavior of Apigenin-7-O-Glucoside from Chrysanthemum Morifolium Tea. Yuxiao Wang et al. Nov 2018.
569) Effect of Different Vegetable Oils on the Flavor of Fried Green Onion (Allium fistulosum L.) Oil. Ruifang Wang et al. Mary 2023.
570) Effect of domestic cooking on human bioavailability of naringenin, chlorogenic acid, lycopene and beta carotene in cherry tomatoes. R Bugianesi et al. Dec 2004.
571) Increases in plasma lycopene concentration after consumption of tomatoes cooked with olive oil. Jeanette M Fielding et al. 2005.
572) Effect of food processing on antioxidants, their bioavailability and potential relevance to human health. Gamze Toydemir et al. June 2022.
573) Effect of food processing on antioxidants, their bioavailability and potential relevance to human health. Gamze Toydemir et al. June 2022.
574) Ultrasonic Treatment Enhances Germination and Affects Antioxidant Gene Expression in Soybean (Glycine max L. Merr). Ayoob Obaid Alfalahi et al. Oct 2022.
575) Effect of food processing on antioxidants, their bioavailability and potential relevance to human health. Gamze Toydemir et al. June 2022.
576) Yuste S., Macià A., Motilva M.-J., Prieto-Diez N., Romero M.-P., Pedret A....Rubió L. Thermal and non-thermal processing of red-fleshed

apple: How are (poly)phenol composition and bioavailability affected? Food & Function. 2020;11(12):10436–10447. doi: 10.1039/D0FO02631J.
577) Maillard reaction in vivo and its relevance to diseases: editorial and dedication. Motoko Takahashi and Naoyuki Taniguchi Apr 2021.
578) Advanced glycation End-products (AGEs): an emerging concern for processed food industries. Chetan Sharma et al. Aug 2015.
579) Advanced glycation End-products (AGEs): an emerging concern for processed food industries. Chetan Sharma et al. Aug 2015.
580) Advanced glycation End-products (AGEs): an emerging concern for processed food industries. Chetan Sharma et al. Aug 2015.
581) The role of advanced glycation end products in aging and metabolic diseases: bridging association and causality. Jyotiska Chaudhuri et al. Sept 2019.
582) Dietary AGEs as Exogenous Boosters of Inflammation. Ma. Eugenia Garay-Sevilla et al. Aug 2021.
583) Heat-induced formation of advanced glycation end-products in ground pork as affected by the addition of acetic acid or citric acid and the storage duration prior to the heat treatments. Hui Lin et al. Oct 2022.
584) Analysis of advanced glycation endproducts in selected food items by ultra-performance liquid chromatography tandem mass spectrometry: Presentation of a dietary AGE database. Jean L J M Scheijen et al. Jan 2016.
585) Advanced glycation End-products (AGEs): an emerging concern for processed food industries. Chetan Sharma et al. Aug 2015.
586) Advanced glycation End-products (AGEs): an emerging concern for processed food industries. Chetan Sharma et al. Aug 2015.
587) Role of Chlorogenic Acids in Controlling Oxidative and Inflammatory Stress Conditions. Ningjian Liang and David D. Kitts et al. Dec 2015.
588) Chlorogenic acid inhibits the formation of advanced glycation end products and associated protein cross-linking. Junghyun Kim et al. Mar 2011.
589) Impact of roasting on the phenolic and volatile compounds in coffee beans. Hanjing Wu et al. Jul 2022.
590) Advanced glycation End-products (AGEs): an emerging concern for processed food industries. Chetan Sharma et al. Aug 2015.
591) Advanced glycation End-products (AGEs): an emerging concern for processed food industries. Chetan Sharma et al. Aug 2015.
592) Advanced glycation End-products (AGEs): an emerging concern for processed food industries. Chetan Sharma et al. Aug 2015.
593) Maillard conjugates and their potential in food and nutritional industries: A review. Xiang Liu, et al. Sept 2020.
594) Potential of resveratrol in mitigating advanced glycation end-products formed in baked milk and baked yogurt. Hang Yu et al. Jul 2020.
595) Potential of resveratrol in mitigating advanced glycation end-products formed in baked milk and baked yogurt. Hang Yu et al. July 2020.
596) Advanced glycation End-products (AGEs): an emerging concern for processed food industries. Chetan Sharma et al. Aug 2015.
597) Nutritional Intake and Biomarker Status in Strict Raw Food Eaters. Klaus Abraham et al. Apr 2023.

598) Intake of Raw Fruits and Vegetables Is Associated With Better Mental Health Than Intake of Processed Fruits and Vegetables. Kate L. Brookie et al. Apr 2018.
599) Advanced glycation End-products (AGEs): an emerging concern for processed food industries. Chetan Sharma et al. Aug 2015.
600) A Systematic Review of Antiaging Effects of 23 Traditional Chinese Medicines. Lixin Wang et al. May 2021.
601) Advanced glycation End-products (AGEs): an emerging concern for processed food industries. Chetan Sharma et al. Aug 2015.
602) Rutin inhibited the advanced glycation end products-stimulated inflammatory response and extra-cellular matrix degeneration via targeting TRAF-6 and BCL-2 proteins in mouse model of osteoarthritis. Xiang Chen et al. Sept 2021.
603) Advanced glycation End-products (AGEs): an emerging concern for processed food industries. Chetan Sharma et al. Aug 2015.
604) Advanced glycation End-products (AGEs): an emerging concern for processed food industries. Chetan Sharma et al. Aug 2015.
605) A Systematic Review of Antiaging Effects of 23 Traditional Chinese Medicines. Lixin Wang et al. May 2021.
606) Advanced glycation End-products (AGEs): an emerging concern for processed food industries. Chetan Sharma et al. Aug 2015.
607) The role of advanced glycation end products in aging and metabolic diseases: bridging association and causality. Jyotiska Chaudhuri et al. Sept 2019.
608) A Review of the Potential Use of Pinene and Linalool as Terpene-Based Medicines for Brain Health: Discovering Novel Therapeutics in the Flavours and Fragrances of Cannabis. Katrina Weston-Green et al. Aug 2021.
609) Metformin and berberine, two versatile drugs in treatment of common metabolic diseases. Haoran Wang et al. Sept 2017.
610) Impacts of metformin and aspirin on life history features and longevity of crickets: trade-offs versus cost-free life extension?. Harvir Hans et al. April 2015.
611) Species of the Genus Salix L.: Biochemical Screening and Molecular Docking Approach to Potential Acetylcholinesterase Inhibitors. by Emilia Gligorić et al. May 2019.
612) Species of the Genus Salix L.: Biochemical Screening and Molecular Docking Approach to Potential Acetylcholinesterase Inhibitors. by Emilia Gligorić et al. May 2019.
613) Metal Toxicity Links to Alzheimer's Disease and Neuroinflammation. Tee Jong Huat, et al. Jan 2019.
614) Solar and geomagnetic activity enhance the effects of air pollutants on atrial fibrillation. Carolina L Zilli Vieira et al. May 2022.
615) Estimation of Daily Ground-Received Global Solar Radiation Using Air Pollutant Data. Xinshuo Zhang et al. April 2022.
616) Dose-dependent detoxication of the airborne pollutant benzene in a randomized trial of broccoli sprout beverage in Qidong, China. Jian-Guo Chen et al. Sept 2019.
617) Dose-dependent detoxication of the airborne pollutant benzene in a

randomized trial of broccoli sprout beverage in Qidong, China. Jian-Guo Chen et al. Sept 2019.
618) New Opportunities to Mitigate the Burden of Disease Caused by Traffic Related Air Pollution: Antioxidant-Rich Diets and Supplements. Jillian Barthelemy et al. Jan 2020.
619) Supplementation of the Diet by Exogenous Myrosinase via Mustard Seeds to Increase the Bioavailability of Sulforaphane in Healthy Human Subjects after the Consumption of Cooked Broccoli. Olukayode Okunade et al. May 2018.
620) New Opportunities to Mitigate the Burden of Disease Caused by Traffic Related Air Pollution: Antioxidant-Rich Diets and Supplements. Jillian Barthelemy et al. Jan 2020.
621) Schisandra chinensis Protects the Skin from Global Pollution by Inflammatory and Redox Balance Pathway Modulations: An In Vitro Study. Edwige Ranouille et al. June 2018.
622) Flavonol-rich dark cocoa significantly decreases plasma endothelin-1 and improves cognition in urban children. Lilian Calderón-Garcidueñas et al. Aug 2013.
623) New Opportunities to Mitigate the Burden of Disease Caused by Traffic Related Air Pollution: Antioxidant-Rich Diets and Supplements. Jillian Barthelemy et al. Jan 2020.
624) Contrasting actions of diesel exhaust particles on the pulmonary and cardiovascular systems and the effects of thymoquinone. Abderrahim Nemmar et al. Dec 2011.
625) Citrus Essential Oils in Aromatherapy: Therapeutic Effects and Mechanisms. Pooja Agarwal et al. Dec 2022.
626) Potent Natural Antioxidant Carveol Attenuates MCAO-Stress Induced Oxidative, Neurodegeneration by Regulating the Nrf-2 Pathway. Imran Malik et al. June 2020.
627) Beneficial of Coriander Leaves (Coriandrum sativum L.) to Reduce Heavy Metals Contamination in Rod Shellfish. Winarti, S. et al. Jan 2018.
628) Effect of Coriandrum sativum L. extract on lead excretion in 3-7 year old children. K. Deldar et al. Oct 2008.
629) Aga M, Iwaki K, Ueda Y, et al. Preventive effect of Coriandrum sativum (Chinese parsley) on localized lead deposition in ICR mice. Journal of Ethnopharmacology. 2001;77(2-3):203–208.
630) Chelation: Harnessing and Enhancing Heavy Metal Detoxification—A Review. Margaret E. Sears et al. Apr 2013.
631) Chelation: Harnessing and Enhancing Heavy Metal Detoxification—A Review. Margaret E. Sears et al. Apr 2013.
632) Possible implementation of salicylate anions in lead detoxification. Chrysoula S. Tzima et al. Jul 2023.
633) Estimation of the Chelating Ability of an Extract from Aronia melanocarpa L. Berries and Its Main Polyphenolic Ingredients Towards Ions of Zinc and Copper.Sylwia Borowska, et al. Mar 2020.
634) The role of solar and geomagnetic activity in endothelial activation and inflammation in the NAS cohort. Jessica E. Schiff, et al. Jul 2022.
635) Association of Circulating C-Reactive Protein and Interleukin-6 with Longevity into the 80s and 90s: The Rancho Bernardo Study. Christina

L. Wassel et al. Oct 2010.
636) Effects of Olive Oil on Markers of Inflammation and Endothelial Functionâ€"A Systematic Review and Meta-Analysis. Lukas Schwingshackl et al. July 2015.
637) Association between high sensitivity C-reactive protein and dietary intake in Vietnamese young women. Ahra Ko,1 et al. Aug 2014.
638) Nutrient Intakes Associated with Elevated Serum C-Reactive Protein Concentrations in Normal to Underweight Breastfeeding Women in Northern Kenya. Masako Fujita et al. Aug 2014.
639) The effect of green tea on C-reactive protein and biomarkers of oxidative stress in patients with type 2 diabetes mellitus: A systematic review and meta-analysis. Omid Asbaghi et al. Aug 2019.
640) Mazidi M, Gao H-K, Rezaie P, Ferns GA.. 2016. The effect of ginger supplementation on serum C-reactive protein, lipid profile and glycaemia: a systematic review and meta-analysis. Food Nutr Res. 60(1):32613.
641) Enhanced Oral Bioavailability of Poorly Absorbed Drugs. I. Screening of Absorption Carrier for the Ceftriaxone Complex. March 2004. Seong-Wan Cho et al. Mar 2004.
642) Roles of methyl jasmonate in improving growth and yield of two varieties of bread wheat (Triticum aestivum) under different irrigation regimes. Zahra Javadipour et al. Aug 2019.
643) Exogenous Application of Glycine Betaine on Sweet Cherry Tree (Prunus avium L.): Effects on Tree Physiology and Leaf Properties. Marta Serapicos et al. Dec 2022.
644) Application of Biostimulants Containing Amino Acids to Tomatoes Could Favor Sustainable Cultivation: Implications for Tyrosine, Lysine, and Methionine. Marina Alfosea-Simón et al. Nov 2020.
645) Soil Management for Vegetable Production on Honeoye Soil with Special Reference to the Use of Hardwood Chips. Free, G. Oct 1971.
646) Salicylic Acid Manipulates Ion Accumulation and Distribution in Favor of Salinity Tolerance in Chenopodium quinoa. Hamid Mohammadi et al. Feb 2022.
647) Roles of methyl jasmonate in improving growth and yield of two varieties of bread wheat (Triticum aestivum) under different irrigation regimes. Zahra Javadipour et al. Aug 2019.
648) Kaolin and Jasmonic acid improved cotton productivity under water stress conditions. Muhammad Nazim et al. Nov 2021.
649) Biostimulant enhances growth and corm production of saffron (Crocus sativus L.) in non-traditional areas of North western Himalayas. Neha Chaudhary et al. Feb 2023.
650) Biostimulant Properties of Seaweed Extracts in Plants: Implications towards Sustainable Crop Production. Omar Ali et al. Mar 2021.
651) Biostimulant Properties of Seaweed Extracts in Plants: Implications towards Sustainable Crop Production. Omar Ali et al. Mar 2021.
652) Saffron extract stimulates growth, improves the antioxidant components of Solanum lycopersicum L., and has an antifungal effect. Amine Khoulati et al. Dec 2019.
653) The effect of herbicides and biostimulants on polyphenol content of potato (Solanum tuberosum L.) tubers and leaves. Krystyna Zarzecka et al.

Jan 2019.
654) Chemical Composition and Antibacterial Activity of the Essential Oil of Centella asiatica. Growing in South Africa. O.A. Oyedeji &A.J. Afolayan. Dec 2004.
655) Chemical Composition of the Essential Oil of Centella asiatica (L.) Urb. from Western Himalayag. Virendra P. Joshi et al. Jan 2007.
656) Chemical Composition of the Essential Oil of Centella asiatica (L.) Urb. from Western Himalayag. Virendra P. Joshi et al. Jan 2007.
657) Developmental study of chemical composition of Centella asiatica (L) Urban as a medicinal plant. Volume 8, Issue 3 - Serial Number 31. September 2016. Pages 13-20.
658) Volatile Needle Terpenoids of Six Pinus Species. Christina Tsitsimpikou eta l. Dec 1999.
659) The cannabinoid receptor 2 agonist, β-caryophyllene, improves working memory and reduces circulating levels of specific proinflammatory cytokines in aged male mice. Lindsey Phillips Lindsey et al. June 2019.
660) Beta-caryophyllene modulates expression of stress response genes and mediates longevity in Caenorhabditis elegans. Aakanksha Pant et al. Sept 2014.
661) Huang, M., Sanchez-Moreiras, A. M., Abel, C., Sohrabi, R., Lee, S., Gershenzon, J., et al. (2012). The major volatile organic compound emitted from Arabidopsis thaliana flowers, the sesquiterpene (E)-β-caryophyllene, is a defense against a bacterial pathogen. New Phytol. 193, 997–1008. doi: 10.1111/j.1469-8137.2011.04001.x
662) Pinus armandii. Wikipedia.
663) Pinus Armandii. Encyclopedia, Science News & Research Reviews.
664) Pinus: Salient Features, Morphology and Reproduction. www.biologyeducare.com.
665) Evaluation of the Inhibitory Effects of Rokumigan and Hachimijiogan on Cataract Formation in a Rat Model of Streptozotocin-induced Type 1 Diabetes Effect of Hachimijiogan on Diabetic Cataracts.
666) The Traditional Japanese Herbal Medicine Hachimijiogan Elicits Neurite Outgrowth Effects in PC12 Cells and Improves Cognitive in AD Model Rats via Phosphorylation of CREB. Kaori Kubota et al. Nov 2017.
667) Successful Treatment of a Senile Diabetic Woman with Cataract with Goshajinkigan. Yoshie Usuki et al. The American Journal of Chinese MedicineVol. 19, No. 03n04, pp. 259-263 (1991)Medicinal Plants Research.
668) Evaluation of the effectiveness of Quinax in the prevention of the development of senile cataract. A. Stankiewicz, Emilia Poppe et al. Aug 1990.
669) Recent Approaches of Ocular Disease and Its Herbal Product Treatment: An Update. Kanchan Butola et al. Apr 2023.
670) Clinical study on evaluation of anti-cataract effect of Triphaladi Ghana Vati and Elaneer Kuzhambu Anjana in Timira (immature cataract). Hitesh Bhati and R Manjusha. Sept 2015.
671) Saffron (Crocus sativus L.) in Ocular Diseases: A Narrative Review of the Existing Evidence from Clinical Studies. Rebekka Heitmar et al. Mar 2019.
672) Use of complementary and alternative therapies by patients with

eye diseases: a hospital-based cross-sectional study from Palestine. www.bmccomplementmedtherapies.biomedcentral.com/articles/10.1186/s12906-020-03188-9.
673) Green tea (Camellia sinensis) protects against selenite-induced oxidative stress in experimental cataractogenesis. S K Gupta et al. Jul 2002.
674) Green tea—a new perspective of glaucoma prevention. Kornelijus Gasiunas1 and Saulius Galgauskas. May 2022.
675) Green Tea Catechins as Therapeutic Antioxidants for Glaucoma Treatment. Tsz Kin Ng et al. June 2023.
676) Safety and Efficacy of Anti-Oxidants and Anti-inflammatory Agents in Glaucoma and Diabetic Retinopathy. The New York Eye & Ear Infirmary. Clinical Trial #NCT02984813.
677) Green Tea Catechins as Therapeutic Antioxidants for Glaucoma Treatment. Tsz Kin Ng et al. June 2023.
678) Neuroprotective role of epigallocatechin-3-gallate in acute glaucoma via the nuclear factor-κB signalling pathway. Wen-Hua Zhang et al. Aug 2021.
679) Preclinical evaluation of anticataract activity of Mentha spicata leaves on isolated goat lens by an in vitro model. Shreya Mohandas et al. Sept 2021.
680) Pharmacological activities of an eye drop containing Matricaria chamomilla and Euphrasia officinalis extracts in UVB-induced oxidative stress and inflammation of human corneal cells. Elisabetta Bigagli et al. Aug 2017.
681) Haji-Ali-Nili N, Khoshzaban F, Karimi M, Rahimi R, Ashrafi E, et al. Effect of a Natural Eye Drop, Made of Plantago Ovata Mucilage on Improvement of Dry Eye Symptoms: A Randomized, Double-blind Clinical Trial. Iran J Pharm Res. 2019;18(3):e126193. https://doi.org/10.22037/ijpr.2019.1100717.
682) Potential Herbs Used in Ocular Diseases. P S Sandhu. et al. Apr 2011.
683) Evaluation And Comparative Study Of Lens Aldose Reductase Inhibitory Activity Of Leaves Extracts Of Merremia Emarginata, Permotrema Perlatum, Tridax Procumbens And Euphorbia Prostrata: Potential For Diabetic Cataract Treatment. Thomson Alex Et Al. 2019. Vol 9 No 2-A (2019): Vol 9, Issue 2-A, 2019 (Special Issue: Seminar On Emerging Trends & Innovations In Pharmaceutical Nanotechnology & Nanomedicine Organised At Cdip, Indore).
684) Sheila Maregesi M, Kauke B, Kagashe G, et al. Traditional Eye Medicines in Tanzania: Products, Health Risk Awareness and Safety Evaluation. Herb Med. 2016, 2:1.
685) Inhibitory effect of two Indian medicinal plants on aldose reductase of rat lens in vitro. Rajesh Kumar et al. Sept 2011.
686) Aldose reductase expression as a risk factor for cataract. Anson Snow, et al. Dec 2014.
687) Preventive effect of onion juice on selenite-induced experimental cataract. Alireza Javadzadeh, et al. May 2009.
688) Effects of some probable antioxidants on selenite-induced cataract formation and oxidative stress-related parameters in rats. H Orhan et al.

Dec 1999.
689) Use of caffeic acid phenethyl ester to prevent sodium-selenite-induced cataract in rat eyes. Selim Doganay et al. Aug 2022.
690) Neuroprotection of retinal cells by Caffeic Acid Phenylethyl Ester (CAPE) is mediated by mitochondrial uncoupling protein UCP2. Mingliang Zhang et al. Dec 2021.
691) Caffeic acid phenethyl ester: A review on its pharmacological importance, and its association with free radicals, COVID-19, and radiotherapy. Seyithan Taysi et al. Dec 2022.
692) Isorhamnetin-3-glucoside alleviates oxidative stress and opacification in selenite cataract in vitro. V Gayathri Devi et al. Sept 2010.
693) Therapeutic Potential of Polar and Non-Polar Extracts of Cyanthillium cinereum In Vitro. Gunjan Guha et al. Jan 2011.
694) Medicinal Plants and Natural Products Used in Cataract Management. Devesh Tewari et al. Jun 2019.
695) Lupeol, a pentacyclic triterpenoid isolated from Vernonia cinerea attenuate selenite induced cataract formation in Sprague Dawley rat pups. Radha Asha et al. Feb 2016.
696) Prevention of selenite-induced cataractogenesis by Origanum vulgare extract.K N Dailami et al. Aug 2010.
697) Emblica officinalis and its enriched tannoids delay streptozotocin-induced diabetic cataract in rats. P Suryanarayana et al. July 2007.
698) Effect of byakangelicin, an aldose reductase inhibitor, on galactosemic cataracts, the polyol contents and Na(+), K(+)ATPase activity in sciatic nerves of strepto-zotocin-induced diabetic rats. K H Shin et al. 1998.
699) Byakangelicin induces cytochrome P450 3A4 expression via transactivation of pregnane X receptors in human hepatocytes. Jian Yang et al. Jan 2011.
700) Byakangelicin. www.pubchem.ncbi.nlm.nih.gov/compound/Byakangelicin
701) Dietary Intake of Coumarins and Furocoumarins through Citrus Beverages: A Detailed Estimation by a HPLC-MS/MS Method Combined with the Linear Retention Index System. Adriana Arigò et al. Jul 2021.
702) Anti-osteoporotic effects of Angelica sinensis (Oliv.) Diels extract on ovariectomized rats and its oral toxicity in rats. Dong Wook Lim and Yun Tai Kim. Oct 2014.
703) Vitex negundo attenuates calpain activation and cataractogenesis in selenite models. B N Rooban et al. Mar 2009.
704) Evaluation of the effect of Spathodea campanulata flower bud exudate on cataractogenesis in rat lenses. Gbemisola Adio Ibidunni et al. Apr 2015.
705) South Korea Patent# KR20210142068A.
706) Effect of tannin on oxidative damage of ocular lens. Y Fukaya et al. 1988.
707) Cataract Preventive Role of Isolated Phytoconstituents: Findings from a Decade of Research. Vuanghao Lim et al. Oct 2018.
708) Isorhamnetin-3-glucoside alleviates oxidative stress and opacification in selenite cataract in vitro. V. Gayathri Devi et al. Sept 2010.

709) Diosmetin protects against retinal injury via reduction of DNA damage and oxidative stress. Zeren Shen et al. Dec 2015.

710) Diosmetin attenuates oxidative stress-induced damage to lens epithelial cells via the mitogen-activated protein kinase (MAPK) pathway. Guanghai Guo and Jin Dong. Apr 2022.

711) The composition of chrysanthemum extracts and their pharmacological functions. Wan-Li LIANG et al (no date listed).

712) Flavonoid Composition of Citrus Juices. Giuseppe Gattuso et al. Aug 2007.

713) The Pharmacological Potential of Rutin. Aditya Ganeshpurkara,b,∗ and Ajay K. Saluja. Apr 2016.

714) Protective effect of buckwheat polyphenols against long-lasting impairment of spatial memory associated with hippocampal neuronal damage in rats subjected to repeated cerebral ischemia. Fengling Pu et al. Apr 2004.

715) www.phenol-explorer.eu. quercetin-3-O-rutinoside.

716) Rutin alleviates colon lesions and regulates gut microbiota in diabetic mice. Cifeng Cai, et al. Mar 2023.

717) Antidiabetic cataract effects of GbE, rutin and quercetin are mediated by the inhibition of oxidative stress and polyol pathway. Qian Lu et al. 2018.

718) Effects of 12-week supplementation of Citrus bergamia extracts-based formulation CitriCholess on cholesterol and body weight in older adults with dyslipidemia: a randomized, double-blind, placebo-controlled trial. Yong Cai et al. Dec 2017.

719) Polyphenols, Antioxidant, Antibacterial, and Biofilm Inhibitory Activities of Peel and Pulp of Citrus medica L., Citrus bergamia, and Citrus medica cv. Salò Cultivated in Southern Italy. Florinda Fratianni et al. Dec 2019.

720) Determination of flavonoids from Cirsium japonicum var. maackii and their inhibitory activities against aldose reductase. Jaemin Lee et al. July 2017.

721) Diosmetin attenuates oxidative stress-induced damage to lens epithelial cells via the mitogen-activated protein kinase (MAPK) pathway. Guanghai Guo and Jin Dong. Spring 2022.

722) Antioxidant Effects and Phytochemical Properties of Seven Taiwanese Cirsium Species Extracts. Zi-Wei Zhao et al. Jul 2021.

723) Neuroprotection in Glaucoma: Old and New Promising Treatments. Dario Rusciano et al. Oct 2017.

724) Combination of a rutin and aescin for treating ocular disorders associated with altered blood circulation, e.g. glaucoma or edema. German Patent #DE10122714A1.

725) Rutin prevents retinal ganglion cell death and exerts protective effects by regulating transforming growth factor-β2/Smad2/3Akt/PTEN signaling in experimental rat glaucoma. Ying Li et al. Apr 2019.

726) Study on anti-cataract drugs from natural sources. II. Effects of buddlejae flos on in vitro aldose reductase activity. H Matsuda et al. Mar 1995.

727) Preventive effect of chlorogenic acid on lens opacity and

cytotoxicity in human lens epithelial cells. Young Sook Kim et al. 2011.
728) Cataract Preventive Role of Isolated Phytoconstituents: Findings from a Decade of Research. Vuanghao Lim et al. Oct 2018.
729) Medicinal Plants and Natural Products Used in Cataract Management. Devesh Tewari,1 et al. June 2019.
730) Phenolic compounds and biological activities of small-size citrus: Kumquat and calamondin. Shyi-Neng Lou and Chi-Tang Ho. Jan 2017.
731) Profiling of Flavonoid and Antioxidant Activity of Fruit Tissues from 27 Chinese Local Citrus Cultivars. Qiyang Chen et al. Feb 2020.
732) Putative Anticataract Properties of Honey Studied by the Action of Flavonoids on a Lens Culture Model. Patricia Vit, Tim John Jacob. 2008.
733) Prevention of in vitro glucose-induced cataract by Vasanjana prepared by Yashtimadhu Kalka (paste of Glycyrrhyza glabra Linn). Manjusha Rajagopala et al. Oct 2021.
734) Potential Properties of Natural Nutraceuticals and Antioxidants in Age-Related Eye Disorders. Jessica Maiuolo, et al. Dec 2022.
735) Clinical study on primary open-angle glaucoma with Ashchyotana, Tarpana and oral medication. Shweta Agrawal and Manjusha Rajagopala. Jan 2017.
736) Effects of D-Limonene on aldose reductase and protein glycation in diabetic rats. Munipally Praveen Kumar et al. Apr 2020.
737) Joglekar MM, Panaskar SN, Chougale AD, Kulkarni MJ, Arvindekar AU, A novel mechanism for antiglycative action of limonene through stabilization of protein conformation, Mol. BioSyst 9 (2013) 2463–2472.
738) Mechanistic targeting of advanced glycation end-products in age-related diseases. Sheldon Rowan et al. Dec 2019.
739) Microwave-superheated Vics Vapo Rub: an ocular public health danger. Anne E Fung 1, Karen W Oxford et al. Feb 2004.
740) Topical ayurvedic ointment-induced chemical injury presenting as bilateral acute keratitis. Pranita Sahay et al. Aug 2017.
741) Elucidation and Regulation of Polyphenols in the Smoking Process of Shanxi Aged Vinegar Sankuan Xie, et al. Jul 2021.
742) Effect of polyphenol and pH on cocoa Maillard-related flavour precursors in a lipidic model system. S. S. Noor-Soffalina et al. Nov 2007. International Journal of Food Science and Technology 2009, 44, 168–180.
743) Effect of malting conditions on phenolic content, Maillard reaction products formation, and antioxidant activity of quinoa seeds. Ramiro Ariel Carciochi, et al. Nov 2016.
744) Biostimulants Improve Plant Growth and Bioactive Compounds of Young Olive Trees under Abiotic Stress Conditions. Giulia Graziani et al. Feb 2022.
745) Reducing lead uptake by plants as a way to lead-free food. Małgorzata Wierzbicka et al. May 2023.
746) The Science Behind Biodynamic Preparations: A Literature Review. Linda Chalker-Scott. Horttechnology • December 2013.
747) NIH researchers identify genetic cause of new vascular disease. Rare disease is first discovered in Undiagnosed Diseases Program. et al. Feb 2011. www.nih.gov/
748) Qigong Exercise Balances Oxygen Supply and Acid-Base to

Modulate Hypoxia: A Perspective Platform toward Preemptive Health & Medicine. Junjie Zhang et al. Mar 2023.
749) Perspectives, Measurability and Effects of Non-Contact Biofield-Based Practices: A Narrative Review of Quantitative Research. Luís Carlos Matos et al. June 2021.
750) Method and apparatus for the treatment of physical and mental disorders with low frequency, low flux density magnetic fields. U.S. Patent #US20100298624A1.
751) Altered States of Consciousness Induced by Exogenous Audio Signals. Asa Young. March 2022.
752) Phase-amplitude coupling at the organism level: The amplitude of spontaneous alpha rhythm fluctuations varies with the phase of the infra-slow gastric basal rhythm. Craig G. Richter et al. Feb 2017.
753) Mechanisms for the Atmosphere-generated Seismic Noise from millihertz to 0.05 Hz. Tanimoto, T et al. Dec 2019.
754) Camellia japonica: A phytochemical perspective and current applications facing its industrial exploitation. Antia G. Pereira, et al. Mar 2022.
755) Chemical composition of essential oil of Pinus halepensis Miller growing in Algeria. Tahar Dob et al. Dec 2005.
756) Chemical composition of essential oil of Pinus halepensis Miller growing in Algeria. Tahar Dob et al. Dec 2005.
757) Antioxidant Potential of Pine Needles: A Systematic Study on the Essential Oils and Extracts of 46 Species of the Genus Pinus. Aikaterini Koutsaviti et al. Jan 2021.
758) Terpenoids as Potential Geroprotectors. Ekaterina Proshkina et al. June 2020.
759) Topical pine tar: History, properties and use as a treatment for common skin conditions. Tanya M Barnes and Kerryn A Greive et al. May 2017.
760) Metabolism of terpenes in the response of grape (Vitis vinifera L.) leaf tissues. to UV-B radiation. Mariana Gil et al. Jan 2012.
761) BenchChem. 3-Carene. www.benchchem.com/product/b045970.
762) Antibacterial and Oxidative Stress-Protective Effects of Five Monoterpenes from Softwood. Riina Muilu-Mäkelä et al. June 2022.
763) Phytochemical analysis of Pinus eldarica bark. S. Iravani and B. Zolfaghar* et al. Jul 2014.
764) Research on the antioxidant, wound healing, and anti-inflammatory activities and the phytochemical composition of maritime pine (Pinus pinaster Ait). İlbrahim Tümen et al. Jan 2018.
765) A randomized controlled trial investigating the effect of Pycnogenol and Bacopa CDRI08 herbal medicines on cognitive, cardiovascular, and biochemical functioning in cognitively healthy elderly people. Con K Stough, et al. Mar 2012.
766) A randomized controlled trial investigating the effect of Pycnogenol and Bacopa CDRI08 herbal medicines on cognitive, cardiovascular, and biochemical functioning in cognitively healthy elderly people. Con K Stough, et al. Mar 2012.
767) Oligopin® Supplementation Mitigates Oxidative Stress in

Postmenopausal Women with Osteopenia: A Randomized, Double-blind, Placebo-Controlled Trial. Ziba Majidi et al. Jan 2021.

768) Woo, J.; Yang, H.; Yoon, M.; Gadhe, C.G.; Pae, A.N.; Cho, S.; Lee, C.J. 3-Carene, a phytoncide from pine tree has a sleep-enhancing effect by targeting the GABA(A)-benzodiazepine receptors. Exp. Neurobiol. 2019, 28, 593–601.

769) Hydroalcoholic extract of needles of Pinus eldarica enhances pentobarbital-induced sleep: possible involvement of GABAergic system. Fatemeh Forouzanfar, et al. Jul 2016.

770) Antibacterial and Oxidative Stress-Protective Effects of Five Monoterpenes from Softwood. Riina Muilu-Mäkelä et al. June 2022.

771) Near Infrared (NIR) Light Therapy of Eye Diseases: A Review. Qin Zhu, et al. Jan 2021.

772) Chemical composition and biological activities of extracts and essential oil of Boswellia dalzielii leaves. Midéko Justin Kohoude et al. Sept 2016.

773) Kimmatkar N, Thawani V, Hingorani L, Khiyani R.. 2003. Efficacy and tolerability of Boswellia serrata extract in treatment of osteoarthritis of knee a randomized double blind placebo controlled trial. Phytomedicine. 10:3–7.

774) Anti-Cancer Potential of Cannabinoids, Terpenes, and Flavonoids Present in Cannabis. Andrea M. Tomko et al. Jul 2020.

775) Antibacterial phenolics from Boswellia dalzielii. Alemika Taiwo E, Onawunmi Grace O and Olugbade, Tiwalade O, Nigerian Journal of Natural Products and Medicines, 2006.

776) Pine oil effects on chemical and thermal injury in mice and cultured mouse dorsal root ganglion neurons. S P Clark et al. Feb 2014.

777) Sox9 is required for determination of the chondrogenic cell lineage in the cranial neural crest. Yuko Mori-Akiyama et al. Aug 2003.

778) Low concentration of 3-carene stimulates the differentiation of mouse osteoblastic MC3T3 E1 subclone 4 cells. Jong-Geun Jeong et al. January 2008.

779) Insecticidal activities of leaf and twig essential oils from Clausena excavata against Aedes aegypti and Aedes albopictus larvae. Sen-Sung Cheng et al. Mar 2009.

780) A Therapeutic Approach for Wound Healing by Using Essential Oils of Cupressus and Juniperus Species Growing in Turkey. Ibrahim Tumen et al. Sept 2011.

781) Chemical Components of Essential Oils From the Leaves of Seven Species Belonging to Rutaceae Family from Binh Chau-Phuoc Buu Nature Reserve. Aug 2020. Institute of Biotechnology.

782) Chemical composition and biological activities of extracts and essential oil of Boswellia dalzielii leaves. Midéko Justin Kohoude et al. Dec 2017.

783) The reproductive system. Robert Tisser and, Rodney Young. 2014. Essential Oil Safety (Second Edition).

784) Skin safety and health prevention: an overview of chemicals in cosmetic products. A. Panico, et al. M Diet-Induced Low-Grade Metabolic Acidosis and Clinical Outcomes: A Review. Renata Alves Carnauba et al.

June 2017.ar 2019.
785) Limonene protects human skin keratinocytes against UVB-induced photodamage and photoaging by activating the Nrf2-dependent antioxidant defense system. K J Senthil Kumar et al. Dec 2022.
786) Skin repair properties of d-Limonene and perillyl alcohol in murine models. Patrizia A d'Alessio et al. Mar 2014.
787) Perillyl Alcohol - Uses, Side Effects, and More. WebMD. www.webmd.com/vitamins/ai/ingredientmono-1161/perillyl-alcohol
788) Higher Dietary Acidity is Associated with Lower Bone Mineral Density in Postmenopausal Iranian Women, Independent of Dietary Calcium Intake. Fatemeh Gholami et al. May 2022.
789) A Comparison of Protein Digestibility Corrected Amino Acid Score with Digestible Indispensable Amino Acid Score to Evaluate Protein Quality of Raw and Roasted Pistachio Nuts.Hannah Bailey et al. June 2020.
790) Believe It or 'Nut': Why It Is Time to Set the Record Straight on Nut Protein Quality. Emma Derbyshire et al. May 2023.
791) Polyamines in Food. N Muñoz-Esparza et al. 2019.
792) Bentonite Clay as a Natural Remedy: A Brief Review. Maryam Moosavi. Sept 2017.
793) Effect of Glycine on Lead Mobilization, Lead-Induced Oxidative Stress, and Hepatic Toxicity in Rats. Yolanda Alcaraz-Contreras et al. July 2011.
794) Low mercury, cadmium and lead concentrations in tuna products from the eastern Pacific. Franklin Isaac Ormaza-González et al. Jul 2020.
795) Evaluation Of The Chelating Effect Of Methanolic Extract Of Coriandrum Sativum And Its Fractions On Wistar Rats Poisoned With Lead Acetate. Miguel Ángel Téllez-López,1 Et Al. Jan 2017.
796) Evaluation Of The Chelating Effect Of Methanolic Extract Of Coriandrum Sativum And Its Fractions On Wistar Rats Poisoned With Lead Acetate. Miguel Ángel Téllez-López Et Al. Jan 2017.
797) Exhausted Yerba Mate Leaves (Ilex paraguariensis) as Biosorbent for the Removal of Metals from Aqueous Solutions. Guillermo J. Copell et al. Dec 2010.
798) Mediterranean diet and red wine protect against oxidative damage in young volunteers. Ines Urquiaga et al. Aug 2010.
799) Vitamin C in Disease Prevention and Cure: An Overview. Shailja Chambial et al. Oct 2013.
800) Biologically Active Compounds in Selected Organic and Conventionally Produced Dried Fruits. Dominika Średnicka-Tober et al. July 2020.
801) The effect of exogenous salicylic acid on antioxidant activity, bioactive compounds and antioxidant system in apricot fruit. Zhen Wang et al. Jan 2015.
802) Yield, Fruit Quality, and Storability of 'Canino' Apricot in Response to Aminoethoxyvinylglycine, Salicylic Acid, and Chitosan. Hayam M. Elmenofy et al. Sept 2021.
803) Effect of oxalic acid and salicylic acid treatments on the post-harvest life of temperate grown apricot varieties (Prunus armeniaca) during controlled atmosphere storage. Mariya Batool et al. Oct 2022.

804) Effects of Apricot Fibre on the Physicochemical Characteristics, the Sensory Properties and Bacterial Viability of Nonfat Probiotic Yoghurts. Oya Berkay Karaca, et al. Jan 2019.
805) Lactic acid fermentation of apricot juice by mono- and mixed cultures of probiotic Lactobacillus and Bifidobacterium strains. Erika Bujna, et al. Apr 2018.
806) Polyphenol content in apricot fruits. H. Gómez-Martínez, et al. Feb 2021.
807) Sensory attribute and antioxidant capacity of Maillard reaction products from enzymatic hydrolysate of bovine bone marrow extract. Nabila Begum et al. May 2020.
808) Effects of Astaxanthin, Lutein, and Zeaxanthin on Eye-Hand Coordination and Smooth-Pursuit Eye Movement after Visual Display Terminal Operation in Healthy Subjects: A Randomized, Double-Blind Placebo-Controlled Intergroup Trial. Keisuke Yoshida et al. Mar 2023.
809) In vitro bioaccessibility of macular xanthophylls from commercial microalgal powders of Arthrospira platensis and Chlorella pyrenoidosa. Cristina Tudor et al. Apr 2021.
810) In vitro bioaccessibility of macular xanthophylls from commercial microalgal powders of Arthrospira platensis and Chlorella pyrenoidosa. Cristina Tudor, et al. Apr 2021.
811) Secret of Longevity Could Be Found in Traditional Japanese Plant That Appears to Slow Aging. Newsweek. Kashmira Gander. Feb 2019.
812) The flavonoid 4,4'-dimethoxychalcone promotes autophagy-dependent longevity across species. Didac Carmona-Gutierrez et al. Feb 2019.
813) Zhang W, Chen H, Ding L, Gong J, Zhang M, Guo W, et al. Trojan horse delivery of 4,4'-dimethoxychalcone for parkinsonian neuroprotection. Adv Sci. (2021) 8:2004555. doi: 10.1002/advs.202004555.
814) Heartwood extract of Acacia catechu induces apoptosis in human breast carcinoma by altering bax/bcl 2 ratio. Nikhil Baban Ghate, et al. Jan 2014.
815) Toxicity study of Bidens pilosa in animals. Yu-Chuan Liang et al. Mar 2020.
816) Bidens pilosa Extract Administered after Symptom Onset Attenuates Glial Activation, Improves Motor Performance, and Prolongs Survival in a Mouse Model of Amyotrophic Lateral Sclerosis. Yasuhiro Kosuge et al. Jan 2020.
817) Effect of the standard herbal preparation, STW5, treatment on dysbiosis induced by dextran sodium sulfate in experimental colitis. Sarah S. Mohamed et al. Jun 2021.
818) STW 5 (Iberogast®) – a safe and effective standard in the treatment of functional gastrointestinal disorders. Bertram Ottillinger et al. Feb 2013.
819) Identification of Proline-Based Diketopiperazines in Roasted Coffee. Michael Ginz and Ulrich H. Engelhardt. July 2000.
820) Inoue S., Takanari J., Abe K., Nagayama A., Ikeya Y., Kohda N. Isolation and structure determination of a heat shock protein inducer, asparagus-derived proline-containing 3-alkyldiketopiperazines (Asparaprolines), from a standardized extract of Asparagus officinalis stem.

Nat. Prod. Commun. 2020;15:1–7. doi: 10.1177/1934578X20914681.
821) Cyclic Dipeptides: The Biological and Structural Landscape with Special Focus on the Anti-Cancer Proline-Based Scaffold. Joanna Bojarska et al. Oct 2021.
822) Isolation and Structure Determination of a Heat Shock Protein Inducer, Asparagus-Derived Proline-Containing 3-Alkyldiketopiperazines (Asparaprolines), From a Standardized Extract of Asparagus officinalis Stem. Shoichiro Inoue et al. Mar 2020.
823) The biofiltration ability of Asparagus densiflorus to remove sulfur dioxide from the indoor atmosphere. Rhiannon de la Rosa* and Mary Savin. Fall 2020.
824) Cyclodextrins and antioxidants. José Manuel López-Nicolás. 2011.
825) Effects of cyclodextrin type on vitamin C, antioxidant activity, and sensory attributes of a mandarin juice enriched with pomegranate and goji berries. Patricia Navarro et al. Jun 2011.
826) The Antioxidant Guaiacol Exerts Fungicidal Activity Against Fungal Growth and Deoxynivalenol Production in Fusarium graminearum. Tao Gao et al. Nov 2021.
827) Guaiacol/β-cyclodextrin for rapid healing of dry socket: antibacterial activity, cytotoxicity, and bone repair-an animal study. Patricia Verónica Aulestia-Viera et al. Mar 2019.
828) Protective Effects of Melon Extracts on Bone Strength, Mineralization, and Metabolism in Rats with Ovariectomy-Induced Osteoporosis. Bongju Kim et al. Aug 2019.
829) Resveratrol Concentration in Muscadine Berries, Juice, Pomace, Purees, Seeds, and Wines. B. J. Ector, et al. 1996.
830) Ecogeographic Conditions Dramatically Affect Trans-Resveratrol and Other Major Phenolics' Levels in Wine at a Semi-Arid Area. Bat-Chen R. Lubin et al. Feb 2022.
831) Cyclodextrins as resveratrol carrier system. Carmen Lucas-Abellán et al. 2007.
832) Cyclodextrins as resveratrol carrier system. Carmen Lucas-Abellán et al. 2007.
833) Enhancement of oral bioavailability of coenzyme Q10 by complexation with γ-cyclodextrin in healthy adults. Keiji Terao. et al. Oct 2006.
834) Bioavailability and Biokinetics of Anthocyanins From Red Grape Juice and Red Wine. Roland Bitsch et al. Dec 2004.
835) Prevalence and Toxicity Characterization of Bacillus cereus in Food Products from Poland. Anna Berthold-Pluta et al. Jul 2019.
836) Ginseng volatile oil prolongs the lifespan and healthspan of Caenorhabditis elegans. Lixin Wang et al. Aug 2022.
837) Antiaging agents: safe interventions to slow aging and healthy life span extension. Ji-Kai Liu. May 2022.
838) Lin C, Zhang X, Su Z, Xiao J, Lv M, Cao Y, Chen Y. Carnosol improved lifespan and healthspan by promoting antioxidant capacity in Caenorhabditis elegans. Oxid Med Cell Longev. 2019;2019:5958043. doi: 10.1155/2019/5958043.
839) Sayed AA. Ferulsinaic acid attenuation of advanced glycation end

products extends the lifespan of Caenorhabditis elegans. J Pharm Pharmacol. 2011;63(3):423–428. doi: 10.1111/j.2042-7158.2010.01222.x.
840) Zheng J, Heber D, Wang M, Gao C, Heymsfield SB, Martin RJ, Greenway FL, Finley JW, Burton JH, Johnson WD, Enright FM, Keenan MJ, Li Z. Pomegranate juice and extract extended lifespan and reduced intestinal fat deposition in Caenorhabditis elegans. Int J Vitam Nutr Res. 2017;87(3–4):149–158. doi: 10.1024/0300-9831/a000570.
841) Volatile composition and sensory quality of Spanish pomegranates (Punica granatum L.). Angel Calín-Sánchez et al. Feb 2011.
842) Upadhyay A, Chompoo J, Taira N, Fukuta M, Tawata S. Significant longevity-extending effects of Alpinia zerumbet leaf extract on the life span of Caenorhabditis elegans. Biosci Biotechnol Biochem. 2013;77(2):217–223. doi: 10.1271/bbb.120351.
843) (Capparis spinosa L. An Updated Review on Its Phytochemistry, Nutritional Value, Traditional Uses, and Therapeutic Potential). Hassan Annaz et al. July 2022.
844) Therapeutic Potential of Quercetin: New Insights and Perspectives for Human Health. Bahare Salehi, et al. May 2020.
845) Accoding to the USDA Database for the Flavonoid Content of Selected Foods, Release 3.1 (2014).
846) www.phenol-explorer.eu. Quercetin
847) Insights into flavor and key influencing factors of Maillard reaction products: A recent update. Shuyun Liu, Hanju Sun et al. Sept 2022.
848) DeLeon, SM; et al. (2018). "DMTS is an effective treatment in both inhalation and injection models for cyanide poisoning using unanesthetized mice". Clinical Toxicology. 56 (25): 332–341. doi:10.1080/15563650.2017.1376749. PMC 6322672. PMID 28922956.
849) Using Rosemary Essential Oil as a Potential Natural Preservative during Stirred-like Yogurt Making. Dalia Gamal Kamel et al. Jul 2022.
850) Nam KH, et al. (2019). "Identification of a novel S6K1 inhibitor, rosmarinic acid methyl ester, for treating cisplatin-resistant cervical cancer." BMC Cancer 19(1): 773.
851) Pre-harvest application of hexanal formulations for improving post-harvest life and quality of mango (Mangifera indica L.) cv. Dashehari. Kirandeep Kaur et al. Nov 2020.
852) Anti-Inflammatory Effect of an Apigenin-Maillard reaction Product in Macrophages and Macrophage-Endothelial Cocultures. Qian Zhou et al. May 2019.
853) Gallic Acid Propyl Ester. Antioxidants Effects in Health. Chapter3.7 - Propyl gallate. Renald Blundell et al. 2022.
854) What is NRF2 and should I use it?. Rolling Meadows Animal Hospital. www.adrianvet.com
855) Longevity-Promoting Pathways and Transcription Factors Respond to and Control Extracellular Matrix Dynamics During Aging and Disease. Tinka Vidović 1 and Collin Y. Ewald. July 2022.
856) The KEAP1-NRF2 System in Healthy Aging and Longevity. Daisuke Matsumaru and Hozumi Motohashi. Dec 2021.
857) The Effect of Aerobic Training and Coriander Seed on Oxidative Stress and Mitochondrial Function Markers in Lung Tissue of Rats Exposed

to H2O2. Zahra Mardani et al. Mar 2021.
858) Coriander leaf extract exerts antioxidant activity and protects against UVB-induced photoaging of skin by regulation of procollagen type I and MMP-1 expression. Eunson Hwang et al. Jul 2014.
859) Effect of pine essential oil and rotating magnetic field on antimicrobial performance. Agata Markowska-Szczupak et al. June 2022.
860) Polyethylene Glycol Improves Function and Reduces Oxidative Stress in Synaptosomal Preparations following Spinal Cord Injury. Jian Luo et al. Aug 2004.
861) The use of polyethylene glycol to reduce the anti-nutritional effects of tannins in goats fed woody species. M Decandia et al. Oct 2000.
862) Combined Magnesium/Polyethylene Glycol Facilitates the Neuroprotective Effects of Magnesium in Traumatic Brain Injury at a Reduced Magnesium Dose. Diana S. Busingye et al. Oct 2016.
863) Caenorhabditis elegans Lipin 1 moderates the lifespan-shortening effects of dietary glucose by maintaining ω-6 polyunsaturated fatty acids. Yoonji Jung et al. June 2020.
864) Natural Compounds and Products from an Anti-Aging Perspective. Geir Bjørklund,1 et al. Oct 2022.
865) Comparative evaluation of hypoglycaemic activity of some Indian medicinal plants in alloxan diabetic rats. Ajit Kar et al. Jan 2003.
866) Chemical inducers of heat shock proteins derived from medicinal plants and cytoprotective genes response. Kanwal Ahmed et al. 2012.
867) Antifeedant Potential of Geranylacetone and Nerylacetone and Their Epoxy-Derivatives against Myzus persicae (Sulz.). Anna Wróblewska-Kurdyk et al. Dec 2022.
868) Protective effect of geranylgeranylacetone, an inducer of heat shock protein 70, against drug-induced lung injury/fibrosis in an animal model. Takayoshi Fujibayash et al. Sept 2009.
869) Neuroprotective effects of geranylgeranylacetone in experimental traumatic brain injury. Zaorui Zhao, et al. Dec 2013.Geranylacetone. National Library of Medicine.
870) www.pubchem.ncbi.nlm.nih.gov/compound/Geranylacetone.
871) Review of Shikonin and Derivatives: Isolation, Chemistry, Biosynthesis, Pharmacology and Toxicology. Snehlata Yadav et al. Jul 2022.
872) Extension of Lifespan in C. elegans by Naphthoquinones That Act through Stress Hormesis Mechanisms. Piper R. Hunt et al. July 2011.
873) Cyanidin-3-O-Glucoside Is An Important Anthocyanin In Several Clones Of Vitis Vinifera L. Pinot Noir Fruits And Resulting Wine From Michigan And New Zealand. Gerard A. Logan et al. Jan 2009.
874) Normal and pathological changes in alpha rhythms. D Samson-Dollfus et al. Jun 1997.
875) Effects of essential oils on central nervous system: Focus on mental health. Lorena R. and Lizarraga-Valderrama. Aug 2020.
876) Study on the Effect of Mentha piperita L. Essential Oil on Electroencephalography upon Stimulation with Different Visual Effects. Shifan Lin et al. Jun 2022.
877) McCraty, R., Atkinson, M., Tomasino, D. and Bradley, R. T., The coherent heart: Heart-brain interactions, psychophysiological coherence,

and the emergence of system-wide order. Integral Review, 2009. 5(2): p. 10-115.

878) McCraty, R., Influence of cardiac afferent input on heart-brain synchronization and cognitive performance. International Journal of Psychophysiology, 2002. 45(1-2): p. 72-73.

879) Boosting Brain Waves Improves Memory.Richard J. Addante Et Al. Nov 2021.

880) Investigation of the Phenolic Component Bioavailability Using the In Vitro Digestion/Caco-2 Cell Model, as well as the Antioxidant Activity in Chinese Red Wine. Chunming Xu et al. Oct 2022.

881) Phenolic acid metabolites of polyphenols act as inductors for hormesis in C. elegans. Benjamin Dilberger et al. Sept 2021.

882) Identification of candidate amino acids involved in the formation of pink-red pigments. Eun Jin Lee et al. Oct 2010.

883) Green Synthesis of Gold Nanoparticles using Aqueous Garlic. Yoki Yulizar et al. Aug 2017

884) Prolongation of life span and improved learning in the senescence accelerated mouse produced by aged garlic extract. T Moriguchi et al. Dec 1994

885) Garlic Consumption and All-Cause Mortality. Xiaoming Shi et al. Jul 2019

886) New insights on low-temperature storage regulating garlic. Rongrong Lu et al. Aug 2023

887) Chemical and Biological Properties of S-1-Propenyl-l-Cysteine in Aged Garlic Extract. Yukihiro Kodera et al. Mar 2017

888) Prolongation of life span and improved learning in the senescence accelerated mouse produced by aged garlic extract. T Moriguchi et al. Dec 1994.

889) Garlic essential oil. Karla Paiva Bocate et al. Jul 2021.

890) Terpenoids as Potential Geroprotectors. Ekaterina Proshkina. Et al. June 2020.

INDEX

Wishing you not only a longlife,but a healthy and prosperous one as well.

Scott Rauvers

Made in United States
Orlando, FL
04 November 2026

90201661R00164